NATEF Standards Job Sheets

Manual Transmissions

Jack Erjavec

THOMSON
DELMAR LEARNING™

Australia Canada Mexico Singapore Spain United Kingdom United States

NATEF Standards Job Sheets

Manual Transmissions

Jack Erjavec

Delmar Staff:

Business Unit Director:
Alar Elken
Executive Editor:
Sandy Clark
Acquisitions Editor:
Sanjeev Rao
Team Assistant:
Matthew Seeley

Executive Marketing Manager:
Maura Theriault
Executive Production Manager:
Mary Ellen Black
Production Manager:
Larry Main
Production Editor:
Betsy Hough

Channel Manager:
Fair Huntoon
Marketing Coordinator:
Brian McGrath
Cover Design:
Michael Egan

Printed in USA
2 3 4 5 6 7 8 9 XXX 05 04 03 02

For more information contact Delmar,
5 Maxwell Drive, PO Box 8007,
Clifton Park, NY 12065-8007.

Or find us on the World Wide Web at http://www.delmar.com or
http://www.autoed.com

ISBN: 0-7668-6369-7

NOTICE TO THE READER

CONTENTS

PREFACE

With every passing day it gets harder to learn all that it takes to become a competent automotive technician. Technological advancements have allowed automobile manufacturers to build safer, more reliable, and more efficient vehicles. This is great for consumers, but along with each advancement comes a need for more knowledge.

Fortunately, students don't need to know it all. In fact, no one person knows everything about everything in an automobile. Although they don't know everything, good technicians do have a solid base of knowledge and skills. The purpose of this book is giving students a chance to develop all the skills and gain the knowledge of a competent technician. It is also the purpose of the guidelines established by the National Automotive Technicians Education Foundation (NATEF).

At the expense of much time and the work of many minds, NATEF has assembled a list of basic tasks for each of its certification areas. These tasks identify the basic skills and knowledge levels of competent technicians. The tasks also identify what is required for a student to start a successful career as a technician.

Most of what this book contains is job sheets. These job sheets relate to the tasks specified by NATEF. The main considerations in the creation of these job sheets were student learning and program certification by NATEF. Students are guided through standard industry-accepted procedures. While they are progressing, students are asked to report their findings and offer their thoughts on the steps they have just completed. The questions asked of the students are thought provoking and require students to apply what they know to what they observe.

The job sheets were designed to be generic; that is, whenever possible, the tasks can be performed on any vehicle from any manufacturer. Completion of the sheets does not require the use of specific brands of tools and equipment; rather, students use what is available. In addition, the job sheets can be used as a supplement to any good textbook.

Words to the Instructor I suggest you grade these job sheets on completion and reasoning. Make sure the students answer all questions, and then look at the reasons to see if the task was actually completed and to get a feel for their understanding of the topic. It'll be easy for students to copy others' measurements and findings, but each student should have his or her own base of understanding, and that will be reflected in the explanations given.

Words to the Student While completing the job sheets, you have a chance to develop the skills you need to be successful. When asked for your thoughts or opinions, think about what you observed. Think about what could have caused those results or conditions. You are not being asked to give accurate explanations for everything you do or everything you observe; you are only asked to think. Thinking leads to understanding. Good technicians are good because they have a basic understanding of what they are doing and of what they are doing it to.

MANUAL TRANSMISSIONS

To prepare you to learn what you should learn from completing the job sheets, some basics must be covered. The discussion begins with an overview of manual transmissions and drivelines, with the emphasis placed on what they do and how they work. This includes the major components and designs of manual transmissions and drivelines and their role in the efficient operation of manual transmissions and drivelines of all designs.

Preparing to work on an automobile would not be complete without addressing certain safety issues. This discussion covers what you should and should not do while working on manual transmissions and drivelines, including the proper ways to deal with hazardous and toxic materials.

NATEF's task list for manual transmissions certification is given, along with definitions of some terms used to describe the tasks. The list gives you a good look at what the experts say you need to know before you can be considered competent to work on manual transmissions and drivelines.

Following the task list are descriptions of the various tools and types of equipment you need to be familiar with. These are the tools you will use to complete the job sheets. They are also the tools NATEF has identified as necessary for servicing manual transmissions and drivelines.

After the tool discussion is a cross-reference guide that shows which NATEF tasks are related to specific job sheets. In most cases, there is a single job sheet for each task. Some tasks are part of a procedure, in which case one job sheet may cover two or more tasks. The remainder of the book contains the job sheets.

BASIC MANUAL TRANSMISSION THEORY

Transmissions or transaxles, drive axles, and differentials perform the important task of manipulating the power produced by the engine and routing it to the driving wheels of the vehicle. Precision-machined and fitted gear sets and shafts change the ratio of speed and power between the engine and drive axles. The flow of power is controlled through a manually operated clutch and shift lever.

Clutch

The clutch (Figure 1) is located between the transmission and the engine, where it provides a mechanical coupling between the engine's flywheel and the transmission's input shaft. The driver operates the clutch through a linkage that extends from the passenger compartment to the bell housing (also called the clutch housing) between the engine and the transmission.

The clutch engages the transmission gradually by allowing a certain amount of slippage between the transmission's input shaft and the flywheel. The components of a clutch assembly are the flywheel, clutch disc, pressure plate assembly, clutch release bearing (or throwout bearing), and the clutch fork.

The flywheel and the pressure plate are the drive members of the clutch. The drive member connected to the transmission input shaft is the clutch disc, also called the friction disc. As long as the clutch is disengaged (clutch pedal depressed), the drive members turn independently of the driven member, and the engine is disconnected from the transmission. However, when the clutch is engaged (clutch

Figure 1 The three main parts of a clutch assembly: the pressure plate, clutch disc, and flywheel.

pedal released), the pressure plate moves against the clutch disc, is squeezed between the two revolving drive members, and is forced to turn at the same speed.

The flywheel is normally made of nodular cast iron, which has a high graphite content to lubricate the engagement of the clutch. Welded to or pressed onto the outside diameter of the flywheel is the starter ring gear. The rear surface of the flywheel is a friction surface, machined very flat to ensure smooth clutch engagement. The flywheel absorbs some of the torsional vibrations of the crankshaft and provides the inertia to rotate the crankshaft through the four strokes.

A bore in the center of the flywheel and crankshaft holds the pilot bushing or bearing, which supports the front end of the transmission's input shaft and keeps it aligned with the engine's crankshaft.

The clutch disc receives the driving motion from the flywheel and pressure plate assembly and transmits that motion to the transmission input shaft. Both sides of the clutch disc are covered with friction material, called friction facings.

Grooves are cut across the face of the friction facings to promote clean disengagement of the driven disc from the flywheel and pressure plate; it also promotes better cooling. The facings are riveted to wave springs, also called cushioning springs, which cause the contact pressure on the facings to rise gradually as the springs flatten out when the clutch is engaged. These springs eliminate chatter when the clutch is engaged and reduce the chance of the clutch disc sticking to the flywheel and pressure plate surfaces when the clutch is disengaged. The wave springs and friction facings are fastened to the steel disc.

The clutch disc is designed to absorb such things as crankshaft vibration, abrupt clutch engagement, and driveline shock. Torsional coil springs allow the disc to rotate slightly in relation to the pressure plate while they absorb the torque forces. The number and tension of these springs is determined by engine torque and vehicle weight. Stop pins limit this torsional movement.

The pressure plate assembly squeezes the clutch disc onto the flywheel with sufficient force to transmit engine torque efficiently. It also must move away from the clutch disc so that the clutch disc can stop rotating, even though the flywheel and pressure plate continue to rotate.

Basically, there are two types of pressure plate assemblies: those with coil springs and those with a diaphragm spring. Both types have a steel cover that bolts to the flywheel and acts as a housing to hold the parts together. In both, there is also the pressure plate, which is a heavy, flat ring made of cast iron. The assemblies differ in the manner in which they move the pressure plate toward and away from the clutch disc.

The pilot bushing or bearing supports the outer end of the transmission's input shaft. This shaft is splined to the clutch disc and transmits power from the engine (when the clutch is engaged) to the transmission. A large bearing in the transmission case supports the transmission end of the input shaft. Because the input shaft extends unsupported from the transmission, a pilot bushing is used to keep it in position. By supporting the shaft, the pilot bushing keeps the clutch disc centered in the pressure plate.

The clutch release bearing, also called a throwout bearing, is usually a sealed, prelubricated ball bearing. Its function is to move the pressure plate release levers or diaphragm spring smoothly and quietly through the engagement and disengagement process.

The release bearing is mounted on an iron casting called a hub, which slides on a hollow shaft at the front of the transmission housing. This hollow shaft is part of the transmission's input shaft bearing retainer.

To disengage the clutch, the release bearing is moved forward on its shaft by the clutch fork. The clutch fork is a forked lever that pivots on a ball stud located in an opening in the bell housing. The forked end slides over the hub of the release bearing and the small end protrudes from the bell housing and connects to the clutch linkage and clutch pedal. The clutch fork moves the release bearing and hub back and forth during engagement and disengagement.

The clutch linkage is a series of parts that connects the clutch pedal to the clutch fork. It is through the clutch linkage that the operator controls the engagement and disengagement of the clutch assembly smoothly and with little effort. Clutch linkage can be mechanical or hydraulic. Mechanical clutch linkages rely on shafts and levers or cables.

Often, the clutch assembly is controlled hydraulically. In a hydraulic linkage, hydraulic fluid pressure transmits motion from one sealed cylinder to another through a hydraulic line. The hydraulic pressure developed by the master cylinder decreases required pedal effort and provides a precise method of controlling clutch operation.

When the clutch pedal is depressed, a pushrod in the clutch master cylinder moves a piston to create hydraulic pressure. This pressure is passed through a line to the clutch slave cylinder. The pressure moves the piston of the slave. Piston travel is transmitted by a pushrod to the clutch fork. The pushrod boot keeps contaminants out of the slave cylinder.

Transmissions

Transmissions and transaxles serve basically the same purpose and operate by the same basic principles. Although the assembly of a transaxle is different than a transmission, the fundamentals and basic components are the same for both. In fact, a transaxle is basically a transmission with the final drive unit housed within the assembly.

A transmission is a system of gears that transfers the engine's power to the drive wheels of the car. The transmission receives torque from the engine through its input shaft when the clutch is engaged. The torque is then transferred through of a set of gears, which either multiply it or transfer it directly. The resultant torque turns the transmission's output shaft, which is indirectly connected to the drive wheels. All transmissions have two primary purposes: they select different speed ratios for a variety of conditions and provide a way to reverse the movement of the vehicle.

Engine torque is applied to the transmission's input shaft when the clutch is engaged. The output shaft (mainshaft) is inserted into the input shaft, but rotates independently of it. The input shaft bearing and a bearing at the rear of the transmission case support the mainshaft. The various speed gears rotate on the mainshaft. Located below or to the side of the input and mainshaft assembly is a countershaft, which is fitted with several sized gears. All these gears, except one, are in constant mesh with the gears on the mainshaft. The remaining gear is in constant mesh with the input gear.

Gear changes occur when a gear is selected by the driver and is locked or connected to the mainshaft. This is accomplished by the movement of a collar that connects the gear to the shaft. Smooth and quiet shifting is possible only when the gears and shaft are rotating at the same speed.

Synchronizers

A synchronizer's primary purpose is to bring components that are rotating at different speeds to one synchronized speed. It also serves to lock these parts together. A single synchronizer is placed between two different speed gears; therefore, transmissions have two or three synchronizer assemblies.

Four types of synchronizers are used in synchromesh transmissions: the block synchronizer, disc and plate synchronizer, plain synchronizer, and pin synchronizer. The most commonly used type on current transmissions is the block type. All synchronizers use friction to synchronize the speed of the gear and shaft before the connection is made.

Block synchronizers consist of a hub, sleeve, blocking ring, and inserts or spring-and-ball detent devices. The synchronizer sleeve surrounds the synchronizer assembly and meshes with the external splines of the hub. The hub is internally splined to the transmission's mainshaft. The outside of the sleeve is grooved to accept the shifting fork. Three slots are equally spaced around the outside of the hub and are fitted with the synchronizer's inserts or spring-and-ball detent assemblies.

The synchronizer's inserts are able to slide back and forth freely in the slots. The inserts are designed with a ridge on their outer surface and insert springs hold this ridge in contact with an internal groove in the synchronizer sleeve. If the synchronizer assembly uses spring-and-ball detents, the balls are held in this groove by their spring. The sleeve is machined to allow it to slide smoothly on the hub.

Bronze or brass blocking rings are positioned at the front and rear of each synchronizer assembly. Blocking rings have notches to accept the insert keys that cause it to rotate at the same speed as the hubs. Around the outside of the blocking ring is a set of beveled dog teeth. These teeth are used for alignment during the shift sequence. The inside of the blocking ring is shaped like a cone, the surface of which has many sharp grooves. These inner surfaces of the blocking rings match the conical shape of the shoulders of the driven gear. These cone-shaped surfaces serve as the frictional surfaces for the synchronizer. The shoulder of the gear also has a ring of beveled dog teeth designed to align with the dog teeth on the blocking ring.

Transmission Designs

All automotive transmissions and transaxles are equipped with a varied number of forward speed gears and one reverse speed. Five-speed transmissions and transaxles are now the most commonly used units. Some of the earlier units were actually four-speeds with an add-on fifth or overdrive gear. Most late-model units incorporate fifth gear in their main assemblies. This is also true of late-model six-speed transmissions and transaxles. The fifth and sixth gears of these units are part of the main gear assembly. Most often each of the two high gears in a six-speed provides an overdrive gear.

Basic Operation

All manual transmissions function in much the same way and have similar parts. The transmission case houses most of the gears in the transmission and has machined surfaces for a cover, rear extension housing, and mounting to the clutch's bell housing. The front bearing retainer is a cast-iron piece bolted to the front of the transmission case.

The counter gear assembly is normally located in the lower portion of the transmission case and is constantly meshed with the input gear. The countershaft contains several different-sized gears that rotate as one solid assembly. Normally the counter gear assembly rotates on several rows of roller or needle bearings.

The speed gears are located on the mainshaft but are not fastened to it and rotate freely on the shaft's journals. The gears are constantly in mesh with the counter gears. The outer end of the mainshaft has splines for the slip-joint yoke of the driveshaft.

A reverse gear is not meshed with the counter gear, as the forward gears are; rather, the reverse idler gear is meshed with it. Normally, reverse gear is engaged by sliding it into mesh with the reverse idler gear. The addition of this third gear causes the reverse gear to rotate in the direction opposite to that of the forward gears.

Shift forks move the synchronizer sleeves to engage and disengage gears. Most shift forks have two fingers that ride in the groove on the outside of the sleeve. The forks are bored to fit over shift rails. Tapered pins are commonly used to fasten the shift forks to the rails. Each of the shift rails has shift lugs that the shift lever fits into. These lugs are also fastened to the rails by tapered pins. The shift rails slide back and forth in bores of the transmission case. Each shift rail has notches in which a spring-loaded ball or bullet drops in to give a detent feel to the shift lever and locate the proper position of the shift fork during gear changes. The shift rails also have notches cut in their sides for interlock plates or pins to fit in. Interlock plates prevent the engagement of two gears at the same time. The lower portion of the shifter assembly fits into the shift rail lugs and moves the shift rails for gear selection.

The ends of the shafts are fitted with large roller or ball bearings pressed onto the shaft. Some transmissions with long shafts use an intermediate bearing to give added strength to the shaft. Small roller or needle bearings are often used on the countershaft, reverse idler gear shaft, and at the connection of the output shaft to the input shaft.

Gear Shift Linkages

Gear shift linkages are either internal or external to the transmission. Internal linkages are located at the side or top of the transmission housing. The control end of the shifter is mounted inside the transmission, as are all the shift controls. Movement of the shifter moves a shift rail and shift fork toward the desired gear and moves the synchronizer sleeve to lock the speed gear to the shaft.

External linkages use rods that are external to the transmission to act on levers connected to the transmission's internal shift rails.

Transaxles

The transmission section of a transaxle is practically identical to that of rear-wheel-drive (RWD) transmissions. It provides for torque multiplication, allows for gear shifting, and is synchronized. It also uses many of the design and operating principles found in transmissions. However, a transaxle also contains the differential gear sets and the connections for the drive axles.

One of the primary distinctions of a transaxle is the absence of the cluster gear assembly. The input shaft gears drive the output shaft gears directly. The output shaft rotates according to each synchro-activated gear's operating ratio.

Normally, a transaxle has two separate shafts: an input shaft and an output shaft. Engine torque is applied to the input shaft, and the revised torque (due to the transaxle's gearing) rotates the output shaft. Normally, the input shaft is located above and parallel to the output shaft. The main speed gears freewheel around the output shaft unless they are locked to the shaft by synchronizers. The main speed gears are in constant mesh with their mating gears on the input shaft and rotate whenever the input shaft rotates.

Some transaxles are equipped with an additional shaft designed to offset the power flow on the output shaft. Power is transferred from the output shaft to the third shaft. The third shaft is added only when an extremely compact transaxle installation is required. The third shaft in some transaxles is an offset input shaft that receives the engine's power and transmits it to a mainshaft.

Some transaxles use a two-cable assembly to shift the gears. One cable is the transmission selector cable and the other is a shifter cable. The selector cable activates the desired shift fork, and the shifter cable causes the engagement of the desired gear.

A pinion gear is machined onto the end of the transaxle's output shaft. This pinion gear is in constant mesh with the differential ring gear. When the output shaft rotates, the pinion gear causes the ring gear to rotate. The resultant torque rotates the other differential gears, which in turn rotate the vehicle's drive axles and wheels.

Front-Wheel-Drive Final Drive Units

One major difference between the differential in a RWD car and the differential in a transaxle is power flow. In a RWD differential, the power flow changes 90 degrees between the drive pinion gear and the ring gear. This change in direction is not needed with most front-wheel-drive (FWD) cars. The transverse engine position places the crankshaft so that it is already rotating in the correct direction. Therefore, the purpose of the differential is only to provide torque multiplication and divide the torque between the drive axle shafts so that they can rotate at different speeds.

Four common configurations are used as the final drives on FWD vehicles: helical, planetary, hypoid, and chain drive. The helical, planetary, and chain final drive arrangements are usually found in transversely mounted power trains. Hypoid final drive gear assemblies are normally used with longitudinal power train arrangements.

Front-Wheel-Drive Drive Axles

The differential side gears are connected to inboard constant velocity (CV) joints by splines. The drive axles extend out from each side of the differential to rotate the car's wheels. The axles are made up of three pieces, linked together, to allow the wheels to turn for steering and to move up and down with the suspension. A short stub shaft extends from the differential to the inner CV joint. Connecting the inner CV joint and the outer CV joint is an axle shaft. Extending from the outer CV joint is a short spindle shaft that fits into the hub of the wheels. To keep dirt and moisture out of the CV joints, a neoprene boot is installed over each CV joint assembly.

Rear-Wheel-Drive Drive Axle Shafts and Bearings

RWD vehicles use a single housing to mount the final drive gears and axles. The entire housing is part of the suspension and helps to locate the rear wheels. Located within the hollow horizontal tubes of the axle housing are the axle shafts. Axle shafts are heavy steel bars splined at the inner end to mesh with the axle side gear in the differential. The driving wheel is bolted to the wheel flange at the outer end of the axle shaft.

There are basically three designs of axles: full-floating, three-quarter-floating, and semifloating. The names refer to where the axle bearing is placed in relation to the axle and the housing and how the axle is supported. The bearing of a full-floating axle is located on the outside of the housing. This places all the vehicle's weight on the axle housing and no weight on the axle.

Three-quarter-floating and semifloating axles are supported by bearings located in the housing and thereby carry some of the weight of the vehicle. Most passenger cars are equipped with three-quarter-floating or semifloating axles. Full-floating axles are commonly found on heavy-duty trucks.

Three designs of axle shaft bearings are used on semifloating axles: ball-type, straight-roller, and tapered-roller bearings. The end-to-end movement of the axle is controlled by a C-type retainer on the inner end of the axle shaft or by a bearing retainer and retainer plate at the outer end of the axle shaft.

The drive axles on most new independent rear suspension (IRS) systems use two U- or CV joints per axle to connect the axle to the differential and the wheels. They are also equipped with linkages and control arms to limit camber changes. The axles of an IRS system are much like those of a FWD system. The outer portion of the axle is supported by an upright or locating member, which is also part of the suspension.

Differentials

A differential is a geared mechanism located between the two driving axles. It rotates the driving axles at different speeds when the vehicle is turning a corner. It also allows both axles to turn at the same speed when the vehicle is moving straight. The gear ratio of the differential's ring-and-pinion gear is used to increase torque, which improves driveability.

On RWD vehicles, power from the transmission's output shaft is transferred by the vehicle's drive shaft to the axle assembly through the pinion flange. This flange is the connecting yoke to the rear universal joint. Power then enters the differential assembly on the pinion gear. The pinion teeth engage the ring gear, which is mounted upright at a 90-degree angle to the pinion. Therefore, as the driveshaft turns, so do the pinion and ring gears.

The ring gear is fastened to the differential case, which is supported by two tapered roller bearings and the rear axle housing. Two beveled differential pinion gears and thrust washers are mounted on the differential pinion shaft. Meshed with the differential pinion gears are two axle side gears splined internally to mesh with the external splines on the left and right axle shafts. The pinion gears and side gears form a square of gears inside the differential case.

The differential pinion gears are free to rotate on their own centers and can travel in a circle as the differential case and pinion shaft rotate. The side gears are meshed with the pinion gears and are free to rotate on their own centers. However, since the side gears are mounted at the centerline of the differential case, they do not travel in a circle with the differential case.

When a car is moving straight ahead, both drive wheels are able to rotate at the same speed. Engine power comes in on the pinion gear and rotates the ring gear. The ring gear rotates the differential case and carries around the pinion gears. As a result, all the gears rotate as a single unit. Each side gear rotates at the same speed and in the same plane, as does the case. Each wheel rotates at the same speed because each axle receives the same rotation.

As the vehicle goes around a corner, the inside wheel travels a shorter distance than the outside wheel. The inside wheel must therefore rotate slower than the outside wheel. In this situation, the differential pinion gears will "walk" forward on the slower-turning or inside side gear. As the pinion gears walk around the slower side gear, they drive the other side gear at a greater speed. An equal percentage of speed is removed from one axle and given to the other; however, the torque applied to each wheel is equal.

Drive Shafts and Universal Joints

On front-engined RWD cars, the torque at the transmission's output shaft is carried through a driveshaft to the final drive unit.

The driveshaft is a tube with universal joint yokes welded to both ends of it. Some drivelines have two driveshafts and four universal joints and use a center support bearing, which serves as the connecting link between the two halves. Four-wheel-drive (4WD) vehicles use two driveshafts: one to drive the front wheels and the other to drive the rear wheels.

As the rear wheels encounter irregularities in the road, the vehicle's springs compress or expand. This changes the angle of the driveline between the transmission and the rear axle housing. It also changes the distance between the transmission and the differential.

In order for the driveshaft to respond to these constant changes, it is equipped with one or more universal joints, which permit variations in the angle of the shaft, and a slip joint, which permits the effective length of the driveline to change.

A universal joint is basically a double-hinged joint consisting of two Y-shaped yokes, one on the driving or input shaft and the other on the driven or output shaft, plus a cross-shaped unit called the cross. A yoke is used to connect the U-joint to the shaft. The four arms of the cross are fitted with bearings in the ends of the two shaft yokes. The input shaft's yoke causes the cross to rotate, and the two other trunnions of the cross cause the output shaft to rotate. When the two shafts are at an angle to each other, the bearings allow the yokes to swing around on their trunnions with each revolution. This action allows two shafts, at a slight angle to each other, to rotate together.

Universal joints allow the driveshaft to transmit power to the rear axle through varying angles that result from the travel of the rear suspension. There are three common designs of universal joints: single universal joints, retained by either an inside or outside snap ring; coupled universal joints; and universal joints held in the yoke by U-bolts or lock plates.

The slip joint assembly includes the transmission's output shaft, the slip joint itself, a yoke, U-joint, and the driveshaft. The output shaft has external splines, which match the internal splines of the slip joint. The meshing of the splines allows the two shafts to rotate together but permits the ends of the shafts to slide along each other. This sliding motion allows for an effective change in the length of the driveshaft, as the drive axles move toward or away from the car's frame. A U-joint connects the yoke of the slip joint to the driveshaft.

Four-Wheel-Drive Systems

Four-wheel-drive vehicles designed for off-the-road use are normally RWD vehicles equipped with a transfer case, a front driveshaft, and a front differential and drive axles. Many 4WD vehicles use three driveshafts. One short driveshaft connects the output of the transmission to the transfer case. The output from the transfer case is then sent to the front and rear axles through separate driveshafts.

Some high-performance cars are equipped with 4WD to improve the handling characteristics of the car. Many of these cars are FWD models converted to 4WD. Normally, FWD cars are modified by adding a transfer case, a rear driveshaft, and a rear axle with a differential. Although this is the typical modification, some cars are equipped with a center differential in place of the transfer case. This differential unit allows the rear and front wheels to turn at different speeds and with different amounts of torque.

The transfer case is usually mounted to the side or rear of the transmission. When a driveshaft is not used to connect the transmission to the transfer case, a chain or gear drive, within the transfer case, receives the engine's power from the transmission, and transfers it to the driveshafts leading to the front and rear drive axles.

The transfer case itself is constructed similar to a transmission. It uses shift forks to select the operating mode, as well as splines, gears, shims, bearings, and other components found in transmissions. The housing is filled with lubricant that reduces friction on all moving parts. Seals hold the lubricant in the case and prevent leakage around shafts and yokes. Shims set up the proper clearance between the internal components and the case.

An electric switch or shift lever, located in the passenger compartment, controls the transfer case so that power is directed to the axles selected by the driver. Power can typically be directed to all four wheels, two wheels, or none of the wheels. On many vehicles, the driver can also select a low-speed range for extra torque while traveling in very adverse conditions.

The rear drive axle of a 4WD vehicle is identical to those used in two-wheel-drive vehicles. The front drive axle is also like a conventional rear axle, except that it is modified to allow the front wheels to steer. Further modifications are also necessary to adapt the axle to the vehicle's suspension system. The differential units housed in the axle assemblies are similar to those found in a RWD vehicle.

SAFETY

In an automotive repair shop, there is great potential for serious accidents, simply because of the nature of the business and the equipment used. Through carelessness, the automotive repair industry can be one of the most dangerous occupations. However, the chances of being injured while working on a car are close to nil if you learn to work safely and use common sense. Safety is the responsibility of everyone in the shop.

Personal Protection

Some procedures, such as grinding, result in tiny particles of metal and dust being thrown off at very high speeds. These metal and dirt particles can easily get into your eyes, causing scratches or cuts on your eyeball. Pressurized gases and liquids escaping a ruptured hose or hose fitting can spray a great distance. If these chemicals get into your eyes, they can cause blindness. Dirt and sharp bits of corroded metal can easily fall into your eyes while you are working under a vehicle.

Eye protection should be worn whenever you are exposed to these risks. To be safe, you should wear safety glasses whenever you are working in the shop. Some procedures may require that you wear other eye protection in addition to safety glasses; for example, when cleaning parts with a pressurized spray, you should wear a face shield. The face shield not only gives added protection to your eyes, but it also protects the rest of your face.

If chemicals such as battery acid, fuel, or solvents get into your eyes, flush them continuously with clean water. Have someone call a doctor, and get medical help immediately.

Your clothing should be well fitted and comfortable but made with strong material. Loose, baggy clothing can easily get caught in moving parts and machinery. Some technicians prefer to wear coveralls or shop coats to protect their personal clothing. Your work clothing should offer you some protection but should not restrict your movement.

Long hair and loose, hanging jewelry can create the same type of hazard as loose-fitting clothing. They can get caught in moving engine parts and machinery. If you have long hair, tie it back or tuck it under a cap.

Never wear rings, watches, bracelets, or neck chains. These can easily get caught in moving parts and cause serious injury.

Always wear leather or similar material shoes or boots with non-slip soles. Steel-tipped safety shoes can give added protection to your feet. Jogging or basketball shoes, street shoes, and sandals are inappropriate in the shop.

Good hand protection is often overlooked. A scrape, cut, or burn can limit your effectiveness at work for many days. A well-fitted pair of heavy work gloves should be worn during operations such as grinding and welding or when handling high-temperature components. Always wear approved rubber gloves when handling strong and dangerous caustic chemicals.

Many technicians wear thin, surgical-type latex gloves whenever they are working on vehicles. These offer little protection against cuts but do offer protection against disease and grease buildup under and around your fingernails. These gloves are comfortable and are quite inexpensive.

Accidents can be prevented simply by the way you act. Following are some guidelines for working in a shop. This list does not include everything you should or shouldn't do; it merely gives some things to think about.

- Never smoke while working on a vehicle or while working with any machine in the shop.
- Playing around is not fun when it sends someone to the hospital.
- To prevent serious burns, keep your skin away from hot metal parts such as the radiator, exhaust manifold, tailpipe, catalytic converter, and muffler.
- Always disconnect electric engine cooling fans when working around the radiator. Many of these can turn on without warning and can easily chop off a finger or hand. Make sure you reconnect the fan after you have completed your repairs.
- When working with a hydraulic press, make sure the pressure is applied in a safe manner. It is generally wise to stand to the side when operating the press.
- Properly store all parts and tools by putting them away in a place where people will not trip over them. This practice not only cuts down on injuries, it also reduces time wasted looking for a misplaced part or tool.

Work Area Safety

Your entire work area should be kept clean and safe. Any oil, coolant, or grease on the floor can make it slippery. To clean up oil, use commercial oil absorbent. Keep all water off the floor. Water is slippery on smooth floors, and electricity flows well through water. Aisles and walkways should be kept clean and wide enough to move through easily. Make sure the work areas around machines are large enough to operate machines safely.

Gasoline is a highly flammable volatile liquid. Something that is *flammable* catches fire and burns easily. A *volatile* liquid is one that vaporizes very quickly. *Flammable volatile* liquids are potential firebombs. Always keep gasoline or diesel fuel in an approved safety can, and never use gasoline to clean your hands or tools.

Handle all solvents (and any liquids) with care to avoid spillage. Keep all solvent containers closed, except when pouring. Proper ventilation is very important in areas where volatile solvents and chemicals are used. Solvent and other combustible materials must be stored in approved and designated storage cabinets or rooms with adequate ventilation. Never light matches or smoke near flammable solvents and chemicals, including battery acids.

Oily rags should also be stored in an approved metal container. When these oily, greasy, or paint-soaked rags are left lying about or are not stored properly, they can spontaneously combust. *Spontaneous combustion* refers to fire that starts by itself, without a match.

Disconnecting the vehicle's battery before working on the electrical system or before welding can prevent fires caused by a vehicle's electrical system. To disconnect the battery, remove the negative or ground cable from the battery and position it away from the battery.

Know where all the shop's fire extinguishers are located. Fire extinguishers are clearly labeled as to type and types of fire they should be used on. Make sure you use the correct type of extinguisher for the type of fire you are dealing with. A multipurpose dry chemical fire extinguisher puts out ordinary combustibles, flammable liquids, and electrical fires. Never put water on a gasoline fire. The water will just spread the fire—the proper fire extinguisher smothers the flames.

During a fire, never open doors or windows unless it is absolutely necessary; the extra draft will only make the fire worse. Make sure the fire department is contacted before or during your attempt to extinguish a fire.

Tool and Equipment Safety

Careless use of simple hand tools, such as wrenches, screwdrivers, and hammers, causes many shop accidents that could be prevented. Keep all hand tools free of grease and in good condition. Tools that slip can cause cuts and bruises. If a tool slips and falls into a moving part, it can fly out and cause serious injury.

Use the proper tool for the job, and make sure the tool is of professional quality. Using poorly made tools or the wrong tools can damage parts, the tool itself, or you. Never use broken or damaged tools.

Safety around power tools is very important. Serious injury can result from carelessness. Always wear safety glasses when using power tools. If the tool is electrically powered, make sure it is properly grounded. Before using it, check for bare wires or cracks in the insulation. When using electrical power tools, never stand on a wet or damp floor. Never leave a running power tool unattended.

Tools that use compressed air are called *pneumatic tools*. Compressed air is used to inflate tires, apply paint, and drive tools. Compressed air can be dangerous when it is not used properly.

When using compressed air, wear safety glasses or a face shield, or both. Particles of dirt and pieces of metal blown by the high-pressure air can penetrate your skin or get into your eyes.

Before using a compressed air tool, check all hose connections. Always hold an air nozzle or air control device securely when starting or shutting off the compressed air. A loose nozzle can whip suddenly and cause serious injury. Never point an air nozzle at anyone. Never use compressed air to blow dirt from your clothes or hair. Never use compressed air to clean the floor or workbench.

Always be careful when raising a vehicle on a lift or a hoist. Adapters and hoist plates must be positioned correctly to prevent damage to the underbody of the vehicle. There are specific lift points that allow the weight of the vehicle to be supported evenly by the adapters or hoist plates. The correct lift points can be found in the vehicle's service manual. Before operating any lift or hoist, carefully read the operating manual and follow the operating instructions.

Once you know the lift supports are properly positioned under the vehicle, raise the lift until the supports contact the vehicle. Check the supports to make sure they are in full contact with the vehicle. Shake the vehicle to make sure it is securely balanced on the lift, and then raise the lift to the desired working height. Before working under a car, make sure the lift's locking devices are engaged.

A vehicle can be raised off the ground by a hydraulic jack. The jack's lifting pad must be positioned under an area of the vehicle's frame or at one of the manufacturer's recommended lift points. Never place the pad under the floor pan or under steering and suspension components, which are easily damaged by the weight of the vehicle. Always position the jack so the wheels of the vehicle can roll as the vehicle is being raised.

Safety stands, also called jack stands, should be placed under a sturdy chassis member, such as the frame or axle housing, to support the vehicle after it has been raised by a jack. Once the safety stands are in position, the hydraulic pressure in the jack should be released slowly until the weight of the vehicle is on the stands. Never move under a vehicle that is supported only by a hydraulic jack. Rest the vehicle on the safety stands before moving under the vehicle.

Heavy parts of the automobile, such as engines, are removed with chain hoists or cranes. Cranes often are called cherry pickers. To prevent serious injury, chain hoists and cranes must be properly attached to the parts being lifted. Always use bolts with enough strength to support the object being lifted. After you have attached the lifting chain or cable to the part that is being removed, have your instructor check it. Place the chain hoist or crane directly over the assembly, then attach the chain or cable to the hoist.

Cleaning parts is a necessary step in most repair procedures. Always wear the appropriate protection when using chemical, abrasive, and thermal cleaners.

Vehicle Operation

When a customer brings a vehicle in for service, shop personnel should follow certain driving rules to ensure the safety of everyone in the shop. For example, before moving a car into the shop, buckle your safety belt. Make sure that no one is nearby, that the way is clear, and that there are no tools or parts under the car before you start the engine.

Check the brakes before putting the vehicle in gear. Then, drive slowly and carefully in and around the shop.

If the engine must be running while work is done on the car, block the wheels to prevent the car from moving. Place the transmission in park for automatic transmissions or in neutral for manual transmissions. Set the parking (emergency) brake. Never stand directly in front of or behind a running vehicle.

Run the engine only in a well-ventilated area to avoid the danger of poisonous carbon monoxide (CO) in the engine exhaust. CO is an odorless but deadly gas. Most shops have an exhaust ventilation system; always use it. Connect the hose from the vehicle's tailpipe to the intake for the vent system. Make sure the vent system is turned on before running the engine. If the work area does not have an exhaust venting system, use a hose to direct the exhaust out of the building.

HAZARDOUS MATERIALS AND WASTES

A typical shop contains many potential health hazards for those working in it. These hazards can cause injury, sickness, impairment, discomfort, and even death. Here is a short list of the different classes of hazards.

- Chemical hazards are caused by high concentrations of vapors, gases, or solids (in the form of dust).
- Hazardous wastes are substances that are the result of a service.
- Physical hazards include excessive noise, vibration, pressures, and temperatures.
- Ergonomic hazards are conditions that impede normal or proper body position and motion.

There are many government agencies charged with ensuring safe work environments for all workers. These include the Occupational Safety and Health Administration (OSHA), Mine Safety and Health Administration (MSHA), and National Institute for Occupational Safety and Health (NIOSH). These agencies, in addition to state and local governments, have instituted regulations that must be understood and followed. Everyone in a shop is responsible for adhering to these regulations.

An important part of a safe work environment is the employees' knowledge of potential hazards. Right-to-know laws concerning all chemicals protect every employee in the shop. The general intent of right-to-know laws is for employers to provide their employees with a safe working place as it relates to hazardous materials.

All employees must be trained about their rights under the legislation, the nature of the hazardous chemicals in the workplace, and the contents of the labels on the chemicals. All information about each chemical must be posted on Material Safety Data Sheets (MSDS) and must be accessible. The manufacturer of the chemical must give these sheets to its customers on request. They detail the chemical composition and precautionary information for all products that can present a health or safety hazard.

Employees must become familiar with the general uses, protective equipment, accident or spill procedures, and any other information about the safe handling of the hazardous material. This training must be given to employees annually and to new employees as part of their job orientation.

All hazardous material must be properly labeled, indicating what health, fire, or reactivity hazard it poses and what protective equipment is necessary when handling each chemical. The manufacturer of the hazardous materials must provide all warnings and precautionary information, which the user must read and understand before using the material. A list of all hazardous materials used in the shop must be posted for employees to see.

Shops must maintain documentation on the hazardous chemicals in the workplace, proof of training programs, records of accidents or spill incidents, and satisfaction of employee requests for specific chemical information via the MSDS. A general right-to-know compliance procedure manual must be used in the shop.

When handling any hazardous materials or hazardous waste, make sure you follow the required procedures for handling such material. Wear the proper safety equipment listed on the MSDS, which includes the use of approved respirator equipment.

Some of the common hazardous materials that automotive technicians use are cleaning chemicals, fuels (gasoline and diesel), paints and thinners, battery electrolyte (acid), used engine oil, refrigerants, and engine coolant (antifreeze).

Many repair and service procedures generate what are known as hazardous wastes. Dirty solvents and cleaners are good examples of hazardous wastes. A material is classified as a *hazardous waste* if it is on the Environmental Protection Agency (EPA) list of known harmful materials or has one or more of the following characteristics.

- *Ignitability*. A liquid with a flash point below 140°F or a solid that can spontaneously ignite.
- *Corrosivity*. It dissolves metals and other materials or burns the skin.
- *Reactivity*. Any material that reacts violently with water or other materials or releases cyanide gas, hydrogen sulfide gas, or similar gases when exposed to low-pH acid solutions. Included are any materials that generate toxic mists, fumes, vapors, or flammable gases.
- *EP toxicity*. Materials that leach one or more of eight heavy metals in concentrations greater than 100 times primary drinking water standard concentrations.

Complete EPA lists of hazardous wastes can be found in the Code of Federal Regulations. It should be noted that no material is considered hazardous waste until the shop is finished using it and ready to dispose of it.

The following list describes the recommended procedure for dealing with some of the common hazardous wastes. Always follow these and any other mandated procedures.

Oil Recycle oil. Set up equipment, such as a drip table or screen table with a used oil collection bucket, to collect oils dripping off parts. Place drip pans underneath vehicles that are leaking fluids onto the storage area. Do not mix other wastes with used oil, except as allowed by your recycler. Used oil generated by a shop (or oil received from household "do-it-yourself" generators) may be burned on site in a commercial space heater. Used oil also may be burned for energy recovery. Contact state and local authorities to determine requirements and to obtain necessary permits.

Oil filters Drain for at least 24 hours, crush, and recycle used oil filters.

Batteries Recycle batteries by sending them to a reclaimer or back to the distributor. Keeping shipping receipts can demonstrate that you have recycled. Store batteries in a watertight, acid-resistant container. Inspect batteries for cracks and leaks when they come in. Treat a dropped battery as if it were cracked. Acid residue is hazardous because it is corrosive and may contain lead and other toxics. Neutralize spilled acid by using baking soda or lime, and dispose of as hazardous material.

Metal residue from machining Collect metal filings when machining metal parts. Keep separate and recycle if possible. Prevent metal filings from falling into a storm sewer drain.

Refrigerants Recover or recycle refrigerants (or do both) during the service and disposal of motor vehicle air conditioners and refrigeration equipment. It is not allowable to knowingly vent refrigerants to the atmosphere. Recovering or recycling during servicing must be performed by an EPA-certified technician using certified equipment and following specified procedures.

Solvents Replace hazardous chemicals with less toxic alternatives that have equal performance. For example, substitute water-based cleaning solvents for petroleum-based solvent degreasers. To reduce the amount of solvent used when cleaning parts, use a two-stage process (i.e., dirty solvent followed by fresh solvent). Hire a hazardous waste management service to clean and recycle solvents. (Some spent solvents must be disposed of as hazardous waste, unless recycled properly.) Store solvents in closed containers to prevent evaporation. Evaporation of solvents contributes to ozone depletion and smog formation. In addition, the residue from evaporation must be treated as a hazardous waste. Properly label spent solvents and store in drip pans or in diked areas and only with compatible materials.

Containers Cap, label, cover, and properly store above ground and outdoors any liquid containers and small tanks within a diked area and on a paved impermeable surface to prevent spills from running into surface or ground water.

Other solids Store materials such as scrap metal, old machine parts, and worn tires under a roof or tarpaulin to protect them from the elements and to prevent potentially contaminated runoff. Consider recycling tires by retreading them.

Liquid recycling Collect and recycle coolants from radiators. Store transmission fluids, brake fluids, and solvents containing chlorinated hydrocarbons separately, and recycle or dispose of them properly.

Shop towels or rags Keep waste towels in a closed container marked "Contaminated Shop Towels Only." To reduce costs and liabilities associated with disposal of used towels, which can be classified as hazardous wastes, investigate using a laundry service that is able to treat the wastewater generated from cleaning the towels.

Waste storage Always keep hazardous waste separate, properly labeled, and sealed in the recommended containers. The storage area should be covered and may need to be fenced and locked if vandalism could be a problem. Select a licensed hazardous waste hauler after seeking recommendations and reviewing the firm's permits and authorizations.

NATEF TASK LIST FOR MANUAL DRIVE TRAIN AND AXLES

A. Clutch Diagnosis and Repair

- A.1. Diagnose clutch noise, binding, slippage, pulsation, and chatter; determine necessary action. — Priority Rating 1
- A.2. Inspect clutch pedal linkage, cables, automatic adjuster mechanisms, brackets, bushings, pivots, and springs; perform necessary action. — Priority Rating 1
- A.3. Inspect hydraulic clutch slave and master cylinders, lines, and hoses; perform necessary action. — Priority Rating 1
- A.4. Inspect release (throw-out) bearing, lever, and pivot; perform necessary action. — Priority Rating 1
- A.5. Inspect and replace clutch pressure plate assembly and clutch disc. — Priority Rating 1
- A.6. Inspect, remove or replace crankshaft pilot bearing or bushing (as applicable). — Priority Rating 1
- A.7. Inspect flywheel and ring gear for wear and cracks, measure runout; determine necessary action. — Priority Rating 1
- A.8. Inspect engine block, clutch (bell) housing, and transmission/transaxle case mating surfaces; determine necessary action. — Priority Rating 3
- A.9. Measure flywheel-to-block runout and crankshaft endplay; determine necessary action. — Priority Rating 3

B. Transmission/Transaxle Diagnosis and Repair

- B.1. Remove and reinstall transmission/transaxle. — Priority Rating 2
- B.2. Disassemble, clean, and reassemble transmission/transaxle components. — Priority Rating 2

B.3.	Inspect transmission/transaxle case, extension housing, case mating surfaces, bores, bushings, and vents; perform necessary action.	Priority Rating 3
B.4.	Diagnose noise, hard shifting, jumping out of gear, and fluid leakage concerns; determine necessary action.	Priority Rating 3
B.5.	Inspect, adjust, and reinstall shift linkages, brackets, bushings, cables, pivots, and levers.	Priority Rating 3
B.6.	Inspect and reinstall power train mounts.	Priority Rating 3
B.7.	Inspect and replace gaskets, seals, and sealants; inspect sealing surfaces.	Priority Rating 2
B.8.	Remove and replace transaxle final drive.	Priority Rating 3
B.9.	Inspect, adjust and reinstall shift cover, forks, levers, grommets, shafts, sleeves, detent mechanism, interlocks, and springs.	Priority Rating 2
B.10.	Measure endplay or preload (shim or spacer selection procedure) on transmission/transaxle shafts; perform necessary action.	Priority Rating 1
B.11.	Inspect and reinstall synchronizer hub, sleeve, keys (inserts), springs, and blocking rings.	Priority Rating 2
B.12.	Inspect and reinstall speedometer drive gear, driven gear, vehicle speed sensor (VSS), and retainers.	Priority Rating 2
B.13.	Diagnose transaxle final drive assembly noise and vibration concerns; determine necessary action.	Priority Rating 3
B.14.	Remove, inspect, measure, adjust, and reinstall transaxle final drive, pinion gears (spiders), shaft, side gears, side bearings, thrust washers, and case assembly.	Priority Rating 2
B.15.	Inspect lubrication devices (oil pump or slingers); perform necessary action.	Priority Rating 3
B.16.	Inspect, test, and replace transmission/transaxle sensors and switches.	Priority Rating 1

C. Drive shaft and Half Shaft, Universal and Constant-Velocity (CV) Joint Diagnosis and Repair

C.1.	Diagnose constant-velocity (CV) joint noise and vibration concerns; determine necessary action.	Priority Rating 2
C.2.	Diagnose universal joint noise and vibration concerns; perform necessary action.	Priority Rating 2
C.3.	Replace front wheel drive (FWD) front wheel bearing.	Priority Rating 2
C.4.	Inspect, service, and replace shafts, yokes, boots, and CV joints.	Priority Rating 1
C.5.	Inspect, service and replace shaft center support bearings.	Priority Rating 3
C.6.	Check shaft balance; measure shaft runout; measure and adjust driveline angles.	Priority Rating 3

D. Drive Axle Diagnosis and Repair

1.	*Ring and Pinion Gears and Differential Case Assembly*	
D.1.1.	Diagnose noise and vibration concerns; determine necessary action.	Priority Rating 2
D.1.2.	Diagnose fluid leakage concerns; determine necessary action.	Priority Rating 2
D.1.3.	Inspect and replace companion flange and pinion seal; measure companion flange runout.	Priority Rating 2
D.1.4.	Inspect ring gear and measure runout; determine necessary action.	Priority Rating 2
D.1.5.	Remove, inspect, and reinstall drive pinion and ring gear, spacers, sleeves, and bearings.	Priority Rating 2
D.1.6.	Measure and adjust drive pinion depth.	Priority Rating 2
D.1.7.	Measure and adjust drive pinion bearing preload.	Priority Rating 1
D.1.8.	Measure and adjust side bearing preload and ring and pinion gear total backlash and backlash variation on a differential carrier assembly (threaded cup or shim types).	Priority Rating 2
D.1.9.	Check ring and pinion tooth contact patterns; perform necessary action.	Priority Rating 1

D.1.10.	Disassemble, inspect, measure, and adjust or replace differential pinion gears (spiders), shaft, side gears, side bearings, thrust washers, and case.	Priority Rating 2
D.1.11.	Reassemble and reinstall differential case assembly; measure runout; determine necessary action.	Priority Rating 2
2.	*Limited Slip Differential*	
D.2.1.	Diagnose noise, slippage, and chatter concerns; determine necessary action.	Priority Rating 3
D.2.2.	Inspect and flush differential housing; refill with correct lubricant.	Priority Rating 2
D.2.3.	Inspect and reinstall clutch (cone or plate) components.	Priority Rating 3
D.2.4.	Measure rotating torque; determine necessary action.	Priority Rating 3
3.	*Drive Axle Shaft*	
D.3.1.	Diagnose drive axle shafts, bearings, and seals for noise, vibration, and fluid leakage concerns; determine necessary action.	Priority Rating 2
D.3.2.	Inspect and replace drive axle shaft wheel studs.	Priority Rating 3
D.3.3.	Remove and replace drive axle shafts.	Priority Rating 1
D.3.4.	Inspect and replace drive axle shaft seals, bearings, and retainers.	Priority Rating 2
D.3.5.	Measure drive axle flange runout and shaft endplay; determine necessary action.	Priority Rating 2

E. Four-wheel Drive/All-wheel Drive Component Diagnosis and Repair

E.1.	Diagnose noise, vibration, and unusual steering concerns; determine necessary action.	Priority Rating 3
E.2.	Inspect, adjust, and repair shifting controls (mechanical, electrical, and vacuum), bushings, mounts, levers, and brackets.	Priority Rating 3
E.3.	Remove and reinstall transfer case.	Priority Rating 3
E.4.	Disassemble, service, and reassemble transfer case and components.	Priority Rating 3
E.5.	Inspect front-wheel bearings and locking hubs; perform necessary action.	Priority Rating 3
E.6.	Check drive assembly seals and vents; check lube level.	Priority Rating 3
E.7.	Diagnose test, adjust, and replace electrical/electronic components of four-wheel drive systems.	Priority Rating 3

DEFINITION OF TERMS USED IN THE TASK LIST

To clarify the intent of these tasks, NATEF has defined some of the terms used in the task list. For a good understanding of what the task includes, refer to this glossary while reading the task list.

adjust	To bring components to specified operational settings.
assemble (reassemble)	To fit together the components of a device.
check	To verify condition by performing an operational or comparative examination.
clean	To rid components of extraneous matter for the purpose of reconditioning, repairing, measuring, and reassembling.
determine	To establish the procedure to be used to effect the necessary repair.
determine necessary action	Indicates that the diagnostic routine(s) is the primary emphasis of a task. The student is required to perform the diagnostic steps and communicate the diagnostic outcomes and corrective actions required to address the concern or problem. The training program determines the communication method (worksheet, test, verbal communication, or other means deemed appropriate) and whether the corrective procedures for these tasks are actually performed.
diagnose	To locate the root cause or nature of a problem by using the specified procedure.

disassemble To separate a component's parts as a preparation for cleaning, inspection, or service.

fill (refill) To bring fluid level to specified point or volume.

find To locate a particular problem, such as shorts, grounds, or opens in an electrical circuit.

flush To use fluid to clean an internal system.

identify To establish the identity of a vehicle or component before service; to determine the nature or degree of a problem.

inspect (See *check*)

install (reinstall) To place a component in its proper position in a system.

listen To use audible clues in the diagnostic process; to hear the customer's description of a problem.

lubricate To employ the correct procedures and materials in performing the prescribed service.

measure To compare existing dimensions to specified dimensions by the use of calibrated instruments and gauges.

mount To attach or place a tool or component in proper position.

perform To accomplish a procedure in accordance with established methods and standards.

perform necessary action Indicates that the student is to perform the diagnostic routine(s) and perform the corrective action item. If various scenarios (conditions or situations) are presented in a single task, at least one of the scenarios must be accomplished.

reassemble (See *assemble*)

refill (See *fill*)

remove To disconnect and separate a component from a system.

repair To restore a malfunctioning component or system to operating condition.

replace To exchange an unserviceable component for a new or rebuilt component; to reinstall a component.

reset (See *set*)

select To choose the correct part or setting during assembly or adjustment.

service To perform a specified procedure when called for in the owner's or service manual.

set To adjust a variable component to a given, usually initial, specification.

test To verify a condition through the use of meters, gauges, or instruments.

torque To tighten a fastener to specified degree or tightness (in a given order or pattern if multiple fasteners are involved on a single component).

verify To establish that a problem exists after hearing the customer's complaint and performing a preliminary diagnosis.

MANUAL TRANSMISSION TOOLS AND EQUIPMENT

Many different tools and pieces of testing and measuring equipment are used to service transmissions and drivelines. NATEF has identified many of these and determined that a certified manual drive train and axle technician must know what they are and how and when to use them. The tools and equipment listed by NATEF are covered in the following discussion. Also included are the tools and equipment you will use while completing the job sheets. Although you must know and be able to use common hand tools, they are not part of this discussion. You should already know what they are and how to use and care for them.

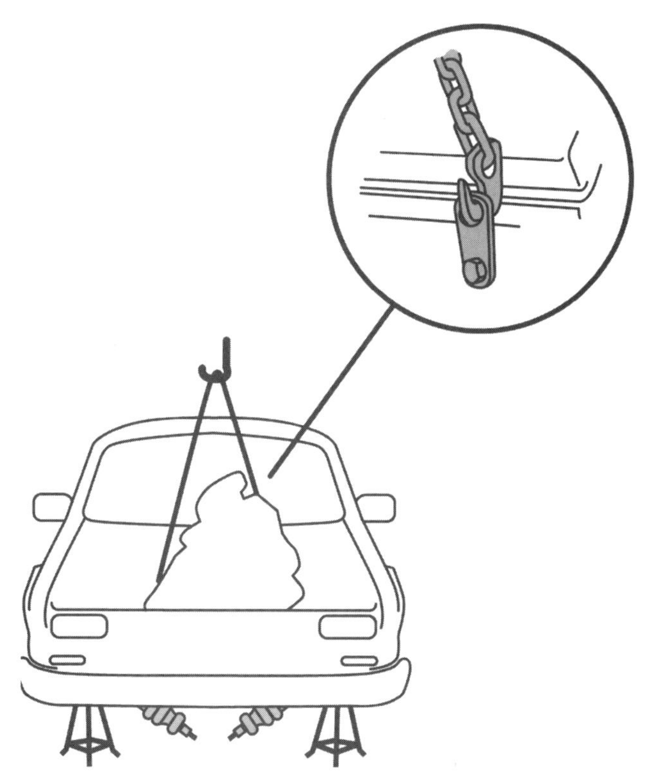

Figure 2 A lifting chain attached to the engine's eye plates.

Portable Crane

To remove and install a transmission, the engine is often moved out of the vehicle with the transmission. To remove or install an engine, a portable crane, frequently called a cherry picker, is used. A crane uses hydraulic pressure that is converted to a mechanical advantage and lifts the engine from the vehicle. To lift an engine, attach a pulling sling or chain to the engine. Some engines have eye plates for use in lifting (Figure 2). If they are not available, the sling must be bolted to the engine. The sling-attaching bolts must be large enough to support the engine and must thread into the block a minimum of 1.5 times the bolt diameter. Connect the crane to the chain. Raise the engine slightly and make sure the sling attachments are secure. Carefully lift the engine out of its compartment.

Lower the engine close to the floor so the transmission can be removed from the engine, if necessary.

Transmission Jacks

Transmission jacks are designed to help you while removing a transmission from under the vehicle. The weight of the transmission makes it difficult and unsafe to remove it without much assistance or a transmission jack. These jacks fit under the transmission (Figure 3) and are typically equipped with hold-down chains. These chains are used to secure the transmission to the jack. The transmission's weight rests on the jack's saddle.

Transmission jacks are available in two basic styles. One is used when the vehicle is raised by a hydraulic jack and is set on jack stands. The other style is used when the vehicle is raised on a lift.

Transaxle Removal and Installation Equipment

Removal and replacement of transversely mounted engines may require special tools. The engines of some FWD vehicles are removed by lifting them from the top. Others must be removed from the bottom, which requires different equipment. Make sure you follow the instructions given by the manufacturer and use the appropriate tools and equipment. The required equipment varies with manufacturer and vehicle model; however, most tools accomplish the same thing.

Figure 3 A typical transmission jack in place under a transmission.

To remove the engine and transmission from under the vehicle, the vehicle must be raised. A crane and/or support fixture (Figure 4) is used to hold the engine and transaxle assembly in place while the assembly is being readied for removal. When everything is set for removal of the assembly, the crane is used to lower the assembly onto a cradle. The cradle is similar to a hydraulic floor jack and is used to lower the assembly further so it can be rolled out from under the vehicle. The transaxle can be separated from the engine once it has been removed from the vehicle.

When the transaxle is removed as a single unit, the engine must be supported while it is in the vehicle before, during, and after transaxle removal. Special fixtures mount to the vehicle's upper frame or suspension parts. These supports have a bracket that is attached to the engine. With the bracket in place, the engine's weight is now on the support fixture, and the transmission can be removed.

Transmission or Transaxle Holding Fixtures

Special holding fixtures should be used to support the transmission or transaxle after it has been removed from the vehicle. These holding fixtures, which may be standalone units or bench mounted, allow the transmission to be repositioned easily during repair work.

Machinist's Rule

A machinist's rule is very much like an ordinary ruler. Each edge of this measuring tool is divided into increments based on a different scale. A typical machinist's rule based on the United States Customary System (USCS) system of measurement may have scales based on 1/8-, 1/16-, 1/32-, and 1/64-inch intervals. Of course, metric machinist rules are also available. Metric rules are usually divided into 0.5-mm and 1-mm increments.

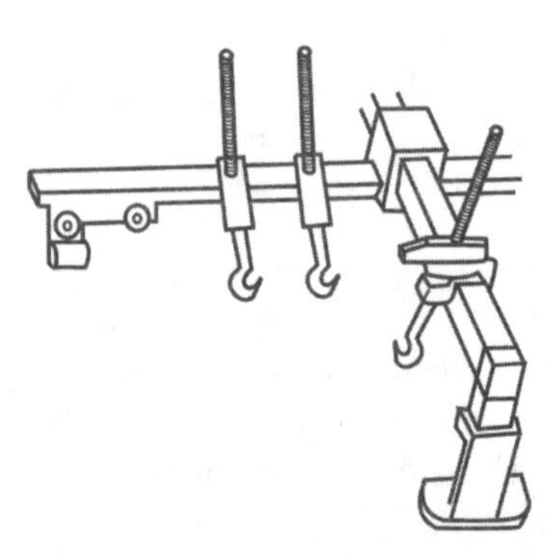

Figure 4 A typical engine support fixture for front-wheel-drive vehicles.

Some machinist's rules are based on decimal intervals. These are typically divided into 1/10-, 1/50-, and 1/1,000-inch (0.1, 0.01, and 0.001) increments. When measuring dimensions that are specified in decimals, decimal machinist's rules are very helpful because you won't need to convert fractions to decimals.

Micrometers

A micrometer is used to measure linear outside and inside dimensions. Both outside and inside micrometers are calibrated and read in the same manner. The major components and markings of a micrometer include the frame, anvil, spindle, lock nut, sleeve, sleeve numbers, sleeve long line, thimble marks, thimble, and ratchet. Micrometers are calibrated in either inch or metric graduations and are available in a range of sizes.

To use and read a micrometer, choose the appropriate size for the object being measured. Typically they measure an inch, therefore the range covered by one-size micrometer measures from 0 to 1 inch, another measures 1 to 2 inches, and so on.

Open the jaws of the micrometer and slip the object between the spindle and the anvil (Figure 5). While holding the object against the anvil, turn the thimble using your thumb and forefinger until the spindle contacts the object. Never clamp the micrometer tightly—use only enough pressure on the thimble to allow the work to just fit between the anvil and spindle. To get accurate readings, you should slip the micrometer back and forth over the object until you feel a very light resistance, while at the same time rocking the tool from side to side to make certain the spindle cannot be closed any further. When a satisfactory adjustment has been made, lock the micrometer. Read the measurement scale.

The graduations on the sleeve each represent 0.025 inch. To read a measurement on a micrometer, begin by counting the visible lines on the sleeve and multiplying them by 0.025. The graduations on the thimble assembly define the area between the lines on the sleeve. The number indicated on the thimble is added to the measurement shown on the sleeve; the sum is the dimension of the object.

Micrometers are available to measure in 0.0001 (ten-thousandths) of an inch. Use this type of micrometer if the specifications call for this much accuracy.

A metric micrometer is read in the same way, except that the graduations are expressed in the metric system of measurement. Each number on the sleeve represents 5 millimeters (mm), or 0.005 meter (m). Each of the 10 equal spaces between each number, with index lines alternating above and below the horizontal line, represents 0.5 mm, or five-tenths of an mm. Therefore, one revolution of the thimble changes the reading one space on the sleeve scale or 0.5 mm. The beveled edge of the thimble is divided into 50 equal divisions with every fifth line numbered: 0, 5, 10, up to 45. Since one complete revolution of the thimble advances the spindle 0.5 mm, each graduation on the thimble is equal to one hundredth of a millimeter. As with the inch-graduated micrometer, the separate readings are added together to obtain the total reading.

Some technicians use a digital micrometer, which is easier to read. These tools do not have the various scales; instead, the measurement is displayed and read directly off the micrometer.

Figure 5 A micrometer is used to measure shim thickness.

Inside micrometers can be used to measure the inside diameter of a bore. To do this, place the tool inside the bore and extend the measuring surfaces until each end touches the bore's surface. If the bore is large, it might be necessary to use an extension rod to increase the micrometer's range. These extension rods come in various lengths. An inside micrometer is read in the same manner as an outside micrometer.

A depth micrometer is used to measure the distance between two parallel surfaces. The sleeves, thimbles, and ratchet screws operate as they do in other micrometers. Depth micrometers are read in the same way as other micrometers.

If a depth micrometer is used with a gauge bar, it is important to keep both the bar and the micrometer from rocking. Any movement of either part results in an inaccurate measurement.

Telescoping Gauge

Telescoping gauges are used for measuring bore diameters and other clearances. They may also be called snap gauges. Telescoping gauges are available in sizes ranging from fractions of an inch through 6 inches. Each gauge consists of two telescoping plungers, a handle, and a lock screw. Snap gauges are normally used with an outside micrometer.

To use a telescoping gauge, insert it into the bore and loosen the lock screw. This will allow the plungers to snap against the bore. Once the plungers have expanded, tighten the lock screw. Then remove the gauge and measure the expanse with a micrometer.

Small Hole Gauge

A small hole or ball gauge works just like a telescoping gauge, but it is designed for small bores. After it is placed into the bore and expanded, it is removed and measured with a micrometer. Like the telescoping gauge, the small hole gauge consists of a lock, a handle, and an expanding end. The end is made to expand or retract by turning the gauge handle.

Feeler Gauge

A feeler gauge is a thin strip of metal or plastic of known and closely controlled thickness. Several of these strips are often assembled together as a feeler gauge set that looks like a pocketknife. The desired thickness gauge can be pivoted away from the others for convenient use. A feeler gauge set usually contains strips or leaves of 0.002- to 0.010-inch thickness (in steps of 0.001 inch) and leaves of 0.012- to 0.024-inch thickness (in steps of 0.002 inch).

A feeler gauge can be used by itself to measure oil pump clearances, gear clearances, endplay, and other distances. It can also be used with a precision straightedge to check the flatness of a sealing surface.

Straightedge

A straightedge is no more than a flat bar machined to be totally flat and straight. To be effective, it must be flat and straight. Any surface that should be flat can be checked with a straightedge and feeler gauge set. The straightedge is placed across and at angles on the surface. At any low points on the surface, a feeler gauge can be placed between the straightedge and the surface. The size gauge that fills in the gap is the amount of warpage or distortion.

Dial Indicator

The dial indicator is calibrated in 0.001-inch (one-thousandth-inch) increments. Metric dial indicators are also available. Both types are used to measure movement. Common uses of the dial indicator include measuring backlash (Figure 6), endplay (Figure 7), and flywheel and axle flange runout.

To use a dial indicator, position the indicator rod against the object to be measured. Push the indicator toward the work until the indicator needle travels far enough around the gauge face to permit movement to be read in either direction. Zero the indicator needle on the gauge. Move the object in the direction required

Figure 6 Checking backlash with a dial indicator.

while observing the needle of the gauge. Always be sure the range of the dial indicator is sufficient to allow the amount of movement required by the measuring procedure. For example, never use a 1-inch indicator on a component that can move 2 inches.

Torque-Indicating Wrench

Torque is the twisting force used to turn a fastener against the friction between threads and between the head of a fastener and the surface of a component. The fact that practically every vehicle and engine manufacturer publishes a list of torque recommendations is ample proof of the importance of using proper amounts of torque when tightening nuts or bolts. The amount of torque applied to a fastener is measured with a torque-indicating, or torque, wrench.

Three basic types of torque-indicating wrenches are available. They are available with pounds-per-inch and pounds-per-foot increments: a beam torque wrench that has a beam pointing to the torque reading; a click-type torque wrench, in which the desired torque reading is set on the handle (when the torque reaches that level, the wrench clicks); and a dial torque wrench, which has a dial that indicates the torque exerted on the wrench. Some designs of this type of torque wrench have a light or buzzer that turns on when the desired torque is reached.

Figure 7 Checking axle shaft endplay with a dial indicator.

Blowgun

Blowguns are used for cleaning parts. Never point a blowgun at yourself or someone else. A blowgun snaps into one end of an air hose and directs airflow when a button is pressed. Always use an OSHA-approved air blowgun. Before using a blowgun, be sure it has not been modified to eliminate air-bleed holes on the side.

Clutch Alignment Tool

To keep the clutch disc centered on the flywheel while assembling the clutch, a clutch alignment tool is used (Figure 8). The tool is inserted through the input shaft opening of the pressure plate and is passed through the clutch disc. The tool is then inserted into the pilot bushing or bearing. The outer diameter (OD) of the alignment tool that goes into the pilot must be only slightly smaller than the inner diameter (ID) of the pilot bushing. The OD of the tool that holds the disc in place also must be only slightly smaller than the ID of the disc's splined bore. The effectiveness of this tool depends on its diameters, which is why it is best to have various sizes of clutch alignment tools.

Gear and Bearing Pullers

Many tools are designed for a specific purpose. An example of a special tool is a gear and bearing puller. Many gears and bearings have a slight interference fit (press fit) when they are installed on a shaft (Figure 9) or in a housing. Something that has a press fit has an interference fit; for example, the ID of a bore is 0.001 inch smaller than the outside diameter of a shaft, so when the shaft is fitted into the bore it must be pressed in to overcome the 0.001-inch interference. This press fit prevents the parts from moving against each other. These gears and bearings must be removed carefully to prevent damage to the gears, bearings, or shafts. Prying or hammering can break or bind the parts. A puller with the proper jaws and adapters should be used to remove gears and bearings. With the proper puller, the force required to remove a gear or bearing can be applied with a slight and steady motion.

Bushing and Seal Pullers and Drivers

Another commonly used group of special tools are the various designs of bushing and seal drivers and pullers. Pullers are either a threaded or slide hammer type of tool. Always make sure you use the correct tool for the job because bushings and seals are easily damaged if the wrong tool or procedure is used. Car manufacturers and specialty tool companies work together closely to design and manufacture special tools for repairing cars. Most of these special tools are listed in the appropriate service manuals.

Figure 8 A clutch alignment tool in place.

Figure 9 Removing side bearings with a threaded puller.

A commonly used bearing puller is a clutch pilot bearing and bushing remover. Typically, the tool is designed to hook into the back of the bearing. Once it is in place, the threaded portion is tightened and the bearing pulled out. The new pilot bearing or bushing is driven into the bore with a hammer and correctly sized driver.

Presses

Many transmission and driveline repairs require the use of a powerful force to assemble or disassemble parts that are press fit together. Removed and installation of axle and final drive bearings (Figure 10), universal joint replacement, and transmission assembly work are just a few of the examples. Presses can be hydraulic, electric, air, or hand driven. Capacities range up to 150 tons of pressing force, depending on the size and design of the press. Smaller arbor and C-frame presses can be bench or pedestal mounted, whereas high-capacity units are freestanding or floor mounted.

Universal Joint Tools

Although universal joints can be serviced with hand tools and a vise, many technicians prefer to use specifically designed tools. One such tool is a C-clamp modified to include a bore that allows the joint's caps to slide in while the clamp is tightened over an assembled joint to remove it. Other tools are the various drivers used with a press to press the joint in and out of its yoke.

Retaining Ring Pliers

Often, a transmission technician will run into many different styles and sizes of retaining rings that hold subassemblies together or keep them in a fixed location. Using the correct tool to remove and

Figure 10 Installing bearings onto a differential case with a driver and hydraulic press.

install these rings is the only safe way to work with them. All transmission and driveline technicians should have an assortment of retaining ring pliers.

Special Tool Sets

Vehicle manufacturers and specialty tool companies work closely together to design and manufacture special tools required to repair transmissions. Most of these special tools are listed in the appropriate service manuals and are part of each manufacturer's essential tool kit.

Service Manuals

Perhaps the most important tools you will use are service manuals. There is no way a technician can remember all the procedures and specifications needed to repair all vehicles. Thus, a good technician relies on service manuals and other information sources for this information. Good information plus knowledge allows a technician to fix a problem with the least frustration and at the lowest expense to the customer.

To obtain the correct transmission specifications and other information, you must first identify the transmission you are working on. The best source for positive identification is the vehicle identification number (VIN). The transmission code can be interpreted through information given in the service manual. The manual may also help you identify the transmission through appearance, casting numbers, and markings on the housing.

The primary source of repair and specification information for any car, van, or truck is the manufacturer. The manufacturer publishes service manuals each year, for every vehicle built. Because of the enormous amount of information, some manufacturers publish more than one manual per year per car model. They are typically divided into sections based on the major systems of the vehicle. In the case of transmissions, there is a section for each transmission that may be in the vehicle. Manufacturers' manuals cover all repairs, adjustments, specifications, detailed diagnostic procedures, and special tools.

Since many technical changes occur on specific vehicles each year, manufacturers' service manuals need to be constantly updated. Updates are published as service bulletins (often referred to as technical service bulletins, or TSBs) that show the changes in specifications and repair procedures during the model year. These changes do not appear in the service manual until the next year. The car manufacturer provides these bulletins to dealers and repair facilities on a regular basis.

Service manuals are also published by independent companies, rather than the manufacturers, but they pay for and get most of their information from the car makers. Service manuals contain component information, diagnostic steps, repair procedures, and specifications for several car makes in one book. Information is usually condensed and is more general in nature than the manufacturer's manuals. The condensed format allows for more coverage in less space and, therefore, is not always specific. They may also contain several years of models as well as several car makes in one book.

Many of the larger parts manufacturers have excellent guides on the various parts they manufacture or supply. They also provide updated service bulletins on their products. Other sources for up-to-date technical information are trade magazines and trade associations.

The same information that is available in service manuals and bulletins is also available on CD-ROMs and DVDs. A single compact disk can hold 250,000 pages of text. This eliminates the need for a huge library to contain all of the printed manuals. Using a CD-ROM to find information is also easier and quicker. The disks are normally updated monthly and contain not only the most recent service bulletins but also engineering and field service fixes.

CROSS-REFERENCE GUIDE

NATEF Task	Job Sheet
A.1	1
A.2	2
A.3	3
A.4	4
A.5	5
A.6	4
A.7	4
A.8	4
A.9	4
B.1	6
B.2	7
B.3	8
B.4	9
B.5	10
B.6	11
B.7	12
B.8	13
B.9	14
B.10	7
B.11	15
B.12	8
B.13	16
B.14	13
B.15	17
B.16	18
C.1	19
C.2	20
C.3	21
C.4	22
C.5	23
C.6	20
D.1.1	16
D.1.2	24
D.1.3	25
D.1.4	26
D.1.5	27
D.1.6	26
D.1.7	26

NATEF Task	Job Sheet
D.1.8	26
D.1.9	26
D.1.10	26
D.1.11	28
D.2.1	29
D.2.2	30
D.2.3	31
D.2.4	29
D.3.1	32
D.3.2	32
D.3.3	32
D.3.4	32
D.3.5	32
E.1	33
E.2	34
E.3	35
E.4	36
E.5	37
E.6	38
E.7	39

JOB SHEETS

MANUAL TRANSMISSIONS JOB SHEET 1

Troubleshoot a Clutch Assembly

Name ______________________ Station ________________ Date ________________

NATEF Correlation

This Job Sheet addresses the following NATEF task:

A.1. Diagnose clutch noise, binding, slippage, pulsation, and chatter; determine necessary action.

Objective

Upon completion of this job sheet, you will be able to demonstrate the ability to troubleshoot a clutch assembly.

Tools and Materials

Droplight
Hoist
Pry bar
Ruler
Service manual
Socket set
Wheel chocks
Wrenches

Protective Clothing

Goggles or safety glasses with side shields

WARNING: *Be sure that the wheels are chocked properly and that the brake system is in good operating condition before beginning this task. Do not allow anyone to stand in front of or behind the vehicle during this test. Do not take any longer than necessary to determine if a problem exists.*

Describe the vehicle being worked on:

Year __________ Make ______________________ Model ______________________

VIN ______________________ Engine type and size ______________________

PROCEDURE (CHECK CLUTCH CHATTER)

1. Start the engine, set the parking brake, depress the clutch pedal fully, and shift the transmission into first gear. Increase the engine speed to about 1500 rpm, and slowly release the clutch pedal. When the pressure plate first makes contact with the clutch disc, notice the clutch operation. Depress the clutch pedal and reduce the engine speed. Record the results on the Report Sheet for Clutch Troubleshooting. ☐ Task completed

2. Shift the transmission into reverse and repeat step 1. Record the results on the Report Sheet for Clutch Troubleshooting. ☐ Task completed

3. If clutch chatter does not occur, increase the engine speed to about 2000 rpm and repeat steps 1 and 2. Record the results on the Report Sheet for Clutch Troubleshooting. ☐ Task completed

4. If chatter occurs during the tests, raise the vehicle on a hoist. Check for loose or broken engine mounts, loose or missing bell housing bolts, and damaged linkage. Record the results on the Report Sheet for Clutch Troubleshooting. Correct any problems found during the inspection. ☐ Task completed

5. Lower the vehicle and repeat steps 1–3. ☐ Task completed

PROCEDURE (CHECK CLUTCH SLIPPAGE)

1. Block the front wheels with wheel chocks and set the parking brake. Start the engine and run it for 15 minutes or until it reaches normal operating temperature. ☐ Task completed

2. Shift the transmission into high gear and increase the engine speed to about 2000 rpm. Release the clutch pedal slowly until the clutch is fully engaged. ☐ Task completed

   CAUTION: *Do not keep the clutch engaged for more than five seconds at a time. The clutch parts could become overheated and be damaged.*

3. If the engine does not stall, raise the vehicle on a hoist and check the clutch linkage. Correct any problems found during the inspection. Record the results on the Report Sheet for Clutch Troubleshooting. ☐ Task completed

4. If any linkage problems were found and corrected, repeat steps 1 and 2. Record any problems in clutch operation on the Report Sheet for Clutch Troubleshooting. ☐ Task completed

PROCEDURE (CHECK CLUTCH DRAG)

NOTE: *The clutch disc and input shaft require about three to five seconds to come to a complete stop after engagement. This is known as "clutch spindown time." This is normal and should not be mistaken for clutch drag.*

1. Start the engine, depress the clutch pedal fully, and shift the transmission into first gear. ☐ Task completed

2. Shift the transmission into neutral, but do not release the clutch pedal. ☐ Task completed
3. Wait 10 seconds. Shift the transmission into reverse. ☐ Task completed
4. If the shift into reverse causes gear clash, raise the vehicle on a hoist. Record the results on the Report Sheet for Clutch Troubleshooting. ☐ Task completed
5. Check the clutch linkage. ☐ Task completed
6. If any linkage problems were found and corrected, repeat steps 1–3. ☐ Task completed

PROCEDURE (CHECK PEDAL PULSATION)

1. Start the engine. Slowly depress the clutch pedal until the clutch just begins to disengage. ☐ Task completed

   NOTE: *A minor pulsation is normal.*

   Depress the clutch pedal further, and check for pulsation as the clutch pedal is depressed to a full stop. Record the results on the Report Sheet for Clutch Troubleshooting.

2. If no rubbing problems are found, remove one drive belt. Start the engine and check for vibration. Shut off the engine. Repeat this process, removing the drive belts and checking for vibration after each drive belt is removed. If the vibration stops after a particular belt is removed, the problem is in the unit driven by that belt. ☐ Task completed

   CAUTION: *Do not run the engine for more than one minute when checking for a vibration with belts removed.*

3. If a vibration is still present, check for a damaged crankshaft vibration damper. Record the results on the Report Sheet for Clutch Troubleshooting. ☐ Task completed

Problems Encountered

___

___

___

Instructor's Comments

___

___

___

Name ______________________ Station ______________ Date ______________

REPORT SHEET FOR CLUTCH TROUBLESHOOTING		
1. Pedal freeplay		
Clutch pedal height specification		
Actual pedal height		
Clutch pedal freeplay specification		
Actual pedal freeplay		
2. Clutch chatter test		
	Yes	*No*
First gear 1500 rpm		
Reverse gear 1500 rpm		
First gear 2000 rpm		
Reverse gear 2000 rpm		
3. Visual inspection		
	Serviceable	*Nonserviceable*
Engine mounts		
Transmission mounts		
Transmission crossover		
Bell housing bolt torque		
Transmission to bell housing bolt torque		
Clutch linkage		
Oil leakage onto disc		

(continued)

Name ______________________ Station ______________ Date ______________

4. Retest for clutch chatter		
	Yes	*No*
First gear 1500 rpm		
Reverse gear 1500 rpm		
First gear 2000 rpm		
Reverse gear 2000 rpm		
5. Clutch slippage test		
Engine stall		
6. Visual inspection		
	Serviceable	*Nonserviceable*
Linkage adjustment		
Linkage components		
Oil leakage		
7. Clutch drag test		
	Yes	*No*
Gear clash		
8. Pedal pulsation test		
	Yes	*No*
Top of pedal travel		
Freeplay removed		
Middle of travel		
End of travel		

Conclusions and Recommendations ____

____

____

MANUAL TRANSMISSIONS JOB SHEET 2

Clutch Linkage Inspection and Service

Name ______________________ Station ______________ Date ______________

NATEF Correlation

This Job Sheet addresses the following NATEF task:

A.2. Inspect clutch pedal linkage, cables, automatic adjuster mechanisms, brackets, bushings, pivots, and springs; perform necessary action.

Objective

Upon completion of this job sheet, you will be able to inspect clutch pedal linkage, cable, automatic adjuster mechanisms, brackets, bushings, pivots, and spring.

Tools and Materials

Lift

Basic hand tools

Protective Clothing

Goggles or safety glasses with side shields

Describe the vehicle being worked on:

Year ________ Make __________________ Model __________________

VIN __________________ Engine type and size __________________

Transmission type and number of forward speeds ______________________

PROCEDURE

1. Road test the vehicle and describe the behavior of the clutch during all observed operations.

____

____

____

____

2. Inspect the following, indicating whether they appear to be okay or not.

	Okay	Should Be Replaced	Not There
Clutch Pedal Linkage	_____	_____	_____
Clutch Cable	_____	_____	_____
Automatic Adjuster Mechanisms	_____	_____	_____
Brackets	_____	_____	_____
Linkage Bushings	_____	_____	_____
Linkage Pivots	_____	_____	_____
Linkage Springs	_____	_____	_____
Clutch Master Cylinder	_____	_____	_____
Clutch Slave Cylinder	_____	_____	_____
Release Lever and Pivot	_____	_____	_____
Power Train Mounts	_____	_____	_____

3. List the problems found in your diagnosis of this clutch.

____

____

____

____

4. List the specifications given in the service manual for the following:

a. Pedal freeplay

____

b. Pedal travel

____

c. Clearance between cable and lever

____

d. Clearance between slave cylinder push rod adjustment nut and wedge

____

e. Torque for adjusting locknut

____

5. Measure the same dimensions on the vehicle.

a. Pedal freeplay

____

b. Pedal travel

____

c. Clearance between cable and lever

____

d. Clearance between slave cylinder push rod adjustment nut and wedge

____

e. Torque for adjusting locknut

____

6. Summarize the results of your measurements.

____

____

7. Check the tightness of the following bolts:
   a. Bell housing to engine block
   b. Transmission to bell housing
   c. Transmission to its mounts
8. Did you need to adjust or tighten any of these to specifications?

____

____

9. Your summary of the above checks.

____

____

____

____

Instructor's Comments

____

____

____

MANUAL TRANSMISSIONS JOB SHEET 3

Inspecting a Hydraulic Clutch Linkage

Name ______________________ Station ______________ Date ______________

NATEF Correlation

This Job Sheet addresses the following NATEF task:

A.3. Inspect hydraulic clutch slave and master cylinders, lines, and hoses; perform necessary action.

Objective

Upon completion of this job sheet, you will be able to inspect the hydraulic lines and hoses for the clutch slave and master cylinder and bleed and adjust the system.

Tools and Materials

Clean fluid for the system

Ruler

Basic hand tools

Protective Clothing

Goggles or safety glasses with side shields

Describe the vehicle being worked on:

Year ________ Make ____________________ Model ____________________

VIN ____________________ Engine type and size ____________________

Transmission type and number of forward speeds ________________________

PROCEDURE

1. With your foot, slowly work the clutch pedal up and down. Pay attention to its entire travel. Check for any binding of the pedal linkage and/or a defective return spring. Describe your findings.

   ____

   ____

2. While working the pedal, does it feel like the hydraulic system is working properly?

   ____

   ____

3. Check the level and the condition of the fluid in the reservoir of the master cylinder. Describe your findings.

   ____

   ____

4. If the fluid level is low, fill the reservoir to the proper level. ☐ Task completed
5. If the fluid was contaminated, the entire hydraulic system must be flushed and bled after a thorough inspection of all of the parts in the system. ☐ Task completed
6. Carefully check the master cylinder, slave cylinder, hydraulic lines and hoses for evidence of leaks. Describe your findings.

   ____

   ____

7. If there was any evidence of a leak, find and repair the leak by replacing the leaking part. ☐ Task completed
8. Anytime the system has been opened to make a repair, the system should be bled. Bleeding may also be necessary if the system was allowed to be very low on fluid. Before bleeding, double-check the system for evidence of leaks and the reservoir for proper fluid level. ☐ Task completed
9. Check all mounting points for the master and slave cylinders. Make sure these do not move or flex when the pedal is depressed. Describe your findings.

   ____

   ____

10. Loosen the bleed screw on the slave cylinder approximately one-half turn. ☐ Task completed
11. Fully depress the clutch pedal, and then move the pedal through three quick and short strokes. ☐ Task completed
12. Close the bleeder screw immediately after the last downward movement of the pedal. Then release the pedal rapidly. ☐ Task completed
13. Recheck and correct the fluid level in the reservoir. ☐ Task completed
14. Repeat steps 10, 11, and 12 until no air is evident in the fluid leaving the bleeder screw. ☐ Task completed
15. Recheck and correct the fluid level in the reservoir. ☐ Task completed
16. Measure clutch pedal free travel according to the manufacturer's recommendations. What were the results of this check?

    ____

    ____

17. Adjust the slave cylinder rod to set the clutch release bearing in its proper location and to set proper pedal free travel ☐ Task completed
18. With your foot, work the clutch pedal and describe how it feels now.

    ____

    ____

    ____

Instructor's Comments

____

____

____

MANUAL TRANSMISSIONS JOB SHEET 4

Clutch Inspection and Service

Name ______________________ Station ______________ Date ______________

NATEF Correlation

This Job Sheet addresses the following NATEF tasks:

A.4. Inspect release (throwout) bearing, lever, and pivot; perform necessary action.

A.6. Inspect, remove, or replace crankshaft pilot bearing or bushing (as applicable).

A.7. Inspect flywheel and ring gear for wear and cracks, measure runout; determine necessary action.

A.8. Inspect engine block, clutch (bell) housing, and transmission/transaxle case mating surfaces; determine necessary action.

A.9. Measure flywheel-to-block runout and crankshaft endplay; determine necessary action.

Objective

Upon completion of this job sheet, you will be able to inspect, measure, and replace the clutch release bearing, pilot bushing or bearing, flywheel, and bell housing.

Tools and Materials

Bell housing alignment tool post assembly
Clutch alignment tool
Dial indicator
Droplight
Feeler gauges
Hoist
Micrometer
Pilot bearing puller
Pry bar
Service manual
Small mirror
Socket set
Straightedge
Surface plate
Torque wrench
Vernier calipers with depth gauge
Workbench

Protective Clothing

Goggles or safety glasses with side shields

Describe the vehicle being worked on:

Year ________ Make ________________ Model ________________

VIN ________________ Engine type and size ________________

PROCEDURE (FOR GENERAL INSPECTION)

1. Perform a visual inspection of the components listed on the report sheet. Check for damage, wear, and warpage. Record your results on the report sheet. ☐ Task completed

2. Locate the specification for flywheel lateral and radial runout on the Report Sheet for Clutch Inspection and Servicing. ☐ Task completed

3. Mount the dial indicator on the bell housing or engine block. Place the tip of the indicator's plunger against the clutch surface of the flywheel. Check the flywheel for excessive lateral and radial runout. Record your results on the Report Sheet for Clutch Inspection and Servicing. ☐ Task completed

4. Locate the specification for flywheel warpage. Record the specification on the Report Sheet for Clutch Inspection and Servicing. ☐ Task completed

5. Using a straightedge and feeler gauge set, check the amount of flywheel warpage and taper. Record your results on the Report Sheet for Clutch Inspection and Servicing. ☐ Task completed

6. If the flywheel has excessive runout, warpage, taper, light scoring, or light checking, remove the flywheel and resurface it. ☐ Task completed

   WARNING: *Do not remove more than 10 percent of the flywheel's original thickness. A flywheel that is too thin can explode, causing serious injury or death.*

PROCEDURE (CHECK FACE RUNOUT)

1. Locate the specification for face runout in the service manual. Record the specification on the Report Sheet for Clutch Inspection and Servicing. ☐ Task completed

2. Attach the clutch assembly to the flywheel. ☐ Task completed

   WARNING: *The clutch assembly is heavy. If it falls, it can damage the assembly, or cause physical injury. Have a helper assist in holding and positioning the clutch assembly during installation.*

3. Install the bell housing alignment tool post assembly through the bell housing bore and into the clutch disc. Tighten the nut on the end of the post assembly until the clamp inside the clutch grips the clutch hub tightly. ☐ Task completed

4. Install the dial indicator on the tool post assembly, and position it so the indicator's plunger contacts the face of the housing in a circular pattern just outside the housing bore. ☐ Task completed

5. Gently pry the crankshaft rearward to eliminate all end play of the crankshaft. Hold the crankshaft in a rearward position. Set the dial indicator to zero. ☐ Task completed

6. Rotate the crankshaft through one complete revolution, and check to see that the dial indicator returns to zero position after one revolution. ☐ Task completed

7. If the dial indicator did not return to zero, repeat steps 2–4. ☐ Task completed

8. If the dial indicator returned to zero, rotate the crankshaft through two revolutions, making note of the highest reading obtained during revolutions. Record your results on the Report Sheet for Clutch Inspection and Servicing. ☐ Task completed

PROCEDURE (CHECK BORE RUNOUT)

1. Position the bell housing alignment tool post assembly and dial indicator to check the bore's runout. ☐ Task completed

   NOTE: *The rubber band should be installed just snug enough to provide a light pressure on the lever tip in the bore of the housing. If the rubber band is too tight, it may bind the dial indicator or distort the readings.*

2. Zero the dial indicator and rotate the crankshaft through one revolution, as in steps 5 and 6 of the procedure for checking face runout. ☐ Task completed

3. Make a note of the highest indicator reading through two complete crankshaft revolutions. Record your results on the Report Sheet for Clutch Inspection and Servicing. ☐ Task completed

Problems Encountered

____

____

____

Instructor's Comments

____

____

____

Name ______________________ Station ______________ Date ______________

REPORT SHEET FOR CLUTCH INSPECTION AND SERVICING		
	Serviceable	*Nonserviceable*
1. Visual Inspection		
Oil leaks		
Flywheel		
Pilot bearing		
Clutch disc		
Pressure plate		
Finger height		
Release bearing		
Clutch fork		
Pressure plate cap screws		
2. Flywheel lateral runout		
Specifications		
Actual		
3. Flywheel radial runout		
Specifications		
Actual		
4. Flywheel warpage		
Specifications		
Actual		
5. Face runout		
Specifications		
Actual		
6. Bore runout		
Specifications		
Actual		

Conclusions and Recommendations ____

____

____

MANUAL TRANSMISSIONS JOB SHEET 5

Clutch Removal, Inspection, and Installation

Name ______________________ Station ________________ Date ________________

NATEF Correlation

This Job Sheet addresses the following NATEF task:

A.5. Inspect and replace clutch pressure plate assembly and clutch disc.

Objective

Upon completion of this job sheet, you will be able to remove, inspect, and replace the clutch pressure plate assembly and clutch disc.

NOTE: *Before the clutch can be removed, both the transmissions/transaxle and the driveline must be removed from the vehicle. Refer to the appropriate job sheets for driveline and transmission/transaxle removal.*

Tools and Materials

Ballpeen hammer
Center punch
Clutch alignment tool
Combination wrenches
Droplight
Fender covers
Flywheel turning tool
Hoist
Jack stands
Service manual
Socket set
Special vacuum cleaner made for removal of asbestos fibers
Transmission jack

Protective Clothing

Goggles or safety glasses with side shields
Respirator and gloves

Describe the vehicle being worked on:

Year __________ Make ______________________ Model ______________________

VIN ______________________ Engine type and size ______________________

PROCEDURE (REMOVE CLUTCH ASSEMBLY)

1. Place fender covers over the fenders. Remove the battery negative cable. ☐ Task completed

   WARNING: *When disconnecting the battery, disconnect the grounded terminal (usually the negative terminal) first. Disconnecting the positive terminal first can create a spark, which can cause the battery to explode and cause serious injury.*

2. Raise the vehicle on a hoist. Remove the starter and driveline. ☐ Task completed

3. Remove the transmission following the appropriate procedures. ☐ Task completed

4. Remove the clutch fork and cross-shaft linkage assemblies. ☐ Task completed

5. Remove the bell housing. There are three basic bell housing configurations commonly used. ☐ Task completed

   a. The bell housing and transmission are removed as a unit. The bell housing may be part of the transmission or bolted to it. ☐ Task completed
   b. The transmission is removed first. The bell housing is unbolted and removed from the back of the engine to expose the clutch assembly. ☐ Task completed
   c. The transmission is removed, but the bell housing remains bolted to the engine. An access opening in the bell housing allows the technician to work on the clutch assembly. ☐ Task completed

6. Remove any dust from the clutch assembly using an OSHA- and EPA-approved vacuum cleaner. ☐ Task completed

WARNING: *The dust inside the bell housing and on the clutch assembly contains asbestos fibers that are a health hazard. Do not blow this dust off with compressed air. Any dust should be removed only with a special vacuum cleaner designed for removal of asbestos fibers.*

7. To ensure proper reassembly, mark the flywheel and pressure plate clutch cover using a hammer and punch. Insert the clutch alignment tool through the clutch disc. Loosen and remove the bolts holding the clutch assembly to the flywheel. ☐ Task completed

CAUTION: *The pressure plate springs will push the pressure plate away from the flywheel. The bolts must be loosened a little at a time. If the bolts are loosened completely, one at a time, the pressure plate springs will push against the flywheel. This causes the clutch cover to become distorted or bent, ruining it.*

WARNING: *The pressure plate assembly is heavy. Be sure the clutch alignment tool remains seated in the pilot bearing. Loosening the mounting bolts may unseat the alignment tool. The alignment tool may have to be tapped in frequently with a soft-faced mallet.*

8. Remove the pressure plate assembly from the vehicle. ☐ Task completed

9. Remove the clutch alignment tool and clutch disc from the flywheel. ☐ Task completed

PROCEDURE (ASSEMBLE CLUTCH)

1. Using the service manual, locate the bolt torque specifications required for the report sheet. Record the specifications on the Report Sheet for Clutch Reassembly and Installation. ☐ Task completed

2. Lubricate the pilot bearing or bushing according to the manufacturer's recommendation. Install the pilot bearing or bushing into the crankshaft flange. ☐ Task completed

3. Install any flywheel dowels. ☐ Task completed

CAUTION: *Do not damage the surface of the flywheel around the dowel holes during installation.*

4. Position the flywheel over the crankshaft mounting flange and screw in, by hand, two mounting bolts to hold the flywheel in position. ☐ Task completed

5. Insert the remaining crankshaft mounting bolts and finger-tighten them. Torque the flywheel mounting bolts in a crisscross pattern to the manufacturer's specifications. ☐ Task completed

6. Place the clutch disc in its position against the flywheel. Line up the matchmarks on the flywheel and clutch cover, and finger-tighten two bolts through the pressure plate assembly to hold it in position. ☐ Task completed

   WARNING: *The clutch assembly is heavy. If it falls, it can damage the assembly, or cause physical injury. Have a helper assist in supporting and positioning the clutch assembly during reassembly.*

7. Select the correct size clutch alignment tool and insert it through the center of the pressure plate and into the clutch disc. Make sure the splines of the disc line up with the splines on the tool if the tool has them. Then carefully push the tool through the disc and into the pilot bearing or bushing. ☐ Task completed

8. Thread the rest of the pressure-plate mounting bolts and finger-tighten them. Torque the bolts in a crisscross pattern. Then remove the alignment tool. ☐ Task completed

   CAUTION: *The clutch assembly-to-flywheel mounting bolts are specially hardened bolts. If it is necessary to replace any of these bolts because of damage, use only the recommended hardened bolts. Do not use ordinary bolts.*

9. Clean the front and rear mounting surfaces of the bellhousing and mounting faces on the rear of engine and front of the transmission. ☐ Task completed

10. Install the bellhousing mounting bolts finger-tight. Torque them to the manufacturer's specifications. ☐ Task completed

11. Press the old throwout bearing out of the clutch fork. Press a new throwout bearing into the clutch fork. ☐ Task completed

12. Slip or clip the new throwout bearing to the clutch fork. ☐ Task completed

13. Put a small amount of the recommended lubricant inside the hub of the release bearing. ☐ Task completed

14. Install the dust cover or seal. ☐ Task completed

15. At this point, the transmission/transaxle and driveline are reinstalled in the vehicle. See the job sheets that pertain to these procedures. After completing all of the reassembly procedures, road test the vehicle and check for proper clutch operation. ☐ Task completed

Problems Encountered

___

___

___

Instructor's Comments

____

____

____

Name ____________________ Station ______________ Date ______________

REPORT SHEET FOR CLUTCH REASSEMBLY AND INSTALLATION		
1. Torque specifications		
Flywheel to crankshaft		
Pressure plate to flywheel		
Bell housing to block		
Transmission to bell housing		
Transmission to mounts		
	Acceptable	*Not Acceptable*
2. Final road test		
Pedal free travel		
Pedal effort		
Pedal pulsation		
Clutch chatter		
Clutch slippage		
Clutch drag		
Vibrations		

Conclusions and Recommendations ____

____

____

MANUAL TRANSMISSIONS JOB SHEET 6

Remove and Install a Transmission or Transaxle

Name ______________________ Station ______________ Date ______________

NATEF Correlation

This Job Sheet addresses the following NATEF task:

B.1. Remove and reinstall a transmission or transaxle.

Objective

Upon completion of this job sheet, you will be able to remove and install a manual transmission or transaxle.

Tools and Materials

Lift
Engine hoist
Engine support
Transmission/transaxle jack
Drain pan

Protective Clothing

Goggles or safety glasses with side shields

Describe the vehicle being worked on:

Year ________ Make ____________________ Model ____________________

VIN ____________________ Engine type and size ____________________

Transmission or Transaxle ____

PROCEDURE

For Removing Transmissions:

1. Disconnect the negative cable of the battery. ☐ Task completed
2. Disconnect the gearshift lever or linkage from the transmission. Describe the procedure you followed to do this:

   ____

   ____

   ____

3. Using chalk, mark the drive shaft at the rear U-joint and the rear axle flange so that the drive shaft can be installed correctly. Remove the drive shaft bolts at the flange of the rear axle. Then remove the drive shaft. ☐ Task completed
4. Disconnect all wiring to the transmission. Mark the wires so you can easily reconnect them to the correct terminals. How did you mark them?

   ____

5. Disconnect the speedometer cable (if equipped). ☐ Task completed

6. Use the transmission jack to support the transmission. ☐ Task completed

7. Loosen and remove the starter motor. If there is enough battery cable to the starter, you may unbolt the starter and move it out of the way without disconnecting the cables. When doing this, use heavy wire to support the starter, and do not allow the starter's weight to pull on the electrical wiring and cables. Describe what you needed to do with the starter.

____

____

8. Loosen and remove the bell housing to engine bolts. ☐ Task completed

9. Slide the transmission away from the engine until the input shaft is away from the clutch assembly. ☐ Task completed

10. Lower the transmission jack and place the transmission in a suitable work place. ☐ Task completed

11. Problems encountered:

____

____

For Removing Transaxles:

1. Disconnect the negative cable of the battery. ☐ Task completed

2. Disconnect the gearshift lever or linkage from the transmission. Describe the procedure you followed to do this:

____

____

____

3. Disconnect all wiring to the transmission. Mark the wires so you can easily reconnect them to the correct terminals. How did you mark them?

____

4. Disconnect the speedometer cable (if equipped). ☐ Task completed

5. Use the transmission jack to support the transmission. ☐ Task completed

6. Loosen and remove the starter motor. If there is enough battery cable to the starter, you may unbolt the starter and move it out of the way without disconnecting the cables. When doing this, use heavy wire to support the starter, and do not allow the starter's weight to pull on the electrical wiring and cables. Describe what you needed to do with the starter.

____

____

7. Identify the transaxle to engine block bolts that cannot be removed from under the vehicle and remove them with the vehicle at ground level. Disconnect anything that may interfere with the movement of the transaxle. Then install the engine support fixture. ☐ Task completed

8. Raise the vehicle and place the transmission jack under the transaxle. ☐ Task completed

9. Remove the suspension parts that will interfere with the removal of the transaxle. Name the parts you needed to remove:

____

____

____

10. Disconnect and remove the drive axles from the transaxle. Describe the procedure you followed to do this:

____

____

____

11. Remove the remaining transaxle to engine bolts. ☐ Task completed

12. Slide the transaxle away from the engine until the input shaft is away from the clutch assembly. ☐ Task completed

13. Problems encountered:

____

____

____

For Installing Transmissions:

1. Place the transaxle on the transmission jack and place the jack into position for installing the transmission into the vehicle. ☐ Task completed

2. Coat the input shaft with a thin layer of grease. Slowly slide the transmission into position. Make sure the splines of the input shaft line up with the internal splines of the clutch disc. If necessary, rotate the input shaft slightly so the splines line up. Do not force the shaft into the disc. Describe any problems you encountered and what adjustments you needed to make to the procedure.

____

____

____

3. Make sure the transmission is fully seated against the engine. Then tighten the bell housing to engine bolts. Is there a torque specification for these bolts? _____ How much? ___________ ☐ Task completed

4. Install the starter motor and its wires and cables. Tighten the bolts evenly. Do not use the bolts to fully seat the starter; instead, hold it in place while tightening the bolts. Some starters use shims to ensure proper tooth contact with the flywheel. If shims are used, use the ones that were removed. Describe what you needed to do with the starter.

____

____

5. Remove the transmission jack. ☐ Task completed
6. Connect the speedometer cable (if equipped). ☐ Task completed
7. Connect all wiring to the transmission according to the marks you made during removal. ☐ Task completed
8. Install the drive shaft so that the index marks made during removal are aligned. Install and tighten the drive shaft bolts at the rear flange. ☐ Task completed
9. Connect the gearshift lever or linkage to the transmission. ☐ Task completed
10. Connect the negative cable of the battery. ☐ Task completed
11. Problems encountered:

____

____

____

For Installing Transaxles:

1. Place the transaxle on the transmission jack and place the jack into position for installing the transaxle into the vehicle. ☐ Task completed
2. Coat the input shaft with a thin layer of grease. Slowly slide the transaxle into position. Make sure the splines of the input shaft line up with the internal splines of the clutch disc. If necessary, rotate the input shaft slightly so the splines line up. Do not force the shaft into the disc. Describe any problems you encountered and what adjustments you needed to make to the procedure.

____

____

____

3. Make sure the transaxle is fully seated against the engine. Then tighten the bell housing to engine bolts. Is there a torque specification for these bolts? _____ How much? ___________
4. Carefully inspect the drive axle boots and install the axles into the transaxle. Describe the procedure you followed to do this:

____

____

____

5. Reinstall the suspension parts that were removed during transaxle removal. Name the parts you needed to reinstall:

____

____

____

6. Remove the transmission jack and lower the vehicle. ☐ Task completed

7. Reinstall and tighten the transaxle to engine bolts that remain out. Remove the engine support fixture. During transaxle removal you may have needed to remove some additional components to allow access to the transaxle or to allow the transaxle to come out. Make sure you reconnect them now. Describe what parts you needed to reinstall.

____

____

____

8. Install the starter motor and wires. Make sure you tighten the bolts evenly and that the starter is fully seated against the engine or transaxle housing. ☐ Task completed

9. Reconnect the speedometer cable (if equipped). ☐ Task completed

10. Reconnect all wiring to the transaxle according to the marks you made during removal. ☐ Task completed

11. Reconnect the gearshift lever or linkage from the transmission. ☐ Task completed

12. Reconnect the negative cable of the battery. ☐ Task completed

13. Problems encountered:

____

____

____

____

Instructor's Comments

____

____

____

MANUAL TRANSMISSIONS JOB SHEET 7

Disassemble and Reassemble a Typical Transaxle

Name ______________________ Station ______________ Date ______________

NATEF Correlation

This Job Sheet addresses the following NATEF tasks:

B.2. Disassemble, clean, and reassemble transmission or transaxle components.

B.10. Measure endplay or preload (shim or spacer selection procedure) on transmission or transaxle shafts; perform necessary action.

Objective

Upon completion of this job sheet, you will be able to disassemble and reassemble a typical manual transaxle. You will also be able to measure endplay or preload (shim or spacer selection procedure) on transmission or transaxle shafts.

Tools and Materials

Basic hand tools

Punch/drift set

Protective Clothing

Goggles or safety glasses with side shields

Describe the vehicle being worked on:

Year ________ Make ____________________ Model ____________________

VIN ____________________ Transaxle type ________________________

PROCEDURE

1. Place the transaxle assembly into a suitable work stand. Secure the appropriate service manual for this transaxle and use it as a guide through the following steps. ☐ Task completed

2. Loosen and remove all transaxle case-to-clutch housing attaching bolts. What did you need to remove in order to do this?

   ____

   ____

3. Separate the housing from the case. If the housing is difficult to loosen, tap it with a soft mallet. ☐ Task completed

4. Remove all external gearshift lever components. To do this what else did you need to remove? Use a punch to drive the roll pin from the shift-lever shaft.

____

____

5. Grasp the input and main shafts and lift them as an assembly from the case. Note the position of the shift forks, so you will know where to place them when reinstalling them. Describe all you needed to do to pull the shafts out.

____

____

6. Remove the differential assembly from the case. Describe all you needed to do to pull the differential case out.

____

____

7. Remove the shift forks. What else did you need to remove?

____

____

8. Begin to remove the bearings and speed gears from the main shaft. These are normally removed in a particular order, describe that order.

____

____

____

9. Inspect the bearings and speed gears. Describe the condition of each:

____

____

____

____

10. Remove the synchronizer assemblies. ☐ Task completed

11. Separate the synchronizer's hub, sleeve, and keys, noting their relative positions and scribing their location on the hub and the sleeve prior to separation. ☐ Task completed

12. Inspect the synchronizer assemblies and describe each part for each assembly:

____

____

____

____

13. Identify all parts that need to be replaced during assembly. List them here:

____

____

____

____

14. Begin reassembly by lightly oiling the parts of the synchronizer. ☐ Task completed
15. Assemble the synchronizer assemblies, being careful to align the index marks made during disassembly. ☐ Task completed
16. Install the synchronizer assemblies onto the main shaft. ☐ Task completed
17. Install the speed gears and bearings onto the main shaft. Did the bearings need to be pressed on?

____

18. Install and tighten the shift fork assemblies. ☐ Task completed
19. Install the differential assembly into the transaxle case. ☐ Task completed
20. Place the main shaft control-shaft assembly on the main shaft so that the shift forks engage in their respective slots in the synchronizer sleeves. Then install the main shaft and input shaft assemblies. ☐ Task completed
21. Install all external gearshift lever components. To do this what else did you need to do?

____

____

22. Properly position and install the shift lever. What else needed to be installed now?

____

____

23. Apply a thin bead of anaerobic sealant on the case's mating surface for the clutch housing. What brand of sealer did you use?

____

24. Install the clutch housing to the case. ☐ Task completed
25. After the housing and case are fit snugly together, tighten the attaching bolts to the specified torque. What is the recommended torque for the bolts?

____

____

____

____

Instructor's Comments

____

____

____

MANUAL TRANSMISSIONS JOB SHEET 8

Inspect and Service an Extension Housing

Name ______________________ Station ______________ Date ______________

NATEF Correlation

This Job Sheet addresses the following NATEF tasks:

B.3. Inspect transmission or transaxle case, extension housing, case mating surfaces, bores, bushings, and vents; perform necessary action.

B.12. Inspect and reinstall speedometer drive gear, driven gear, vehicle speed sensor (VSS), and retainers.

Objective

Upon completion of this job sheet, you will be able to inspect and service the extension housing, its bushing and seals, and the speedometer drive assembly.

Tools and Materials

Hand file

Puller and driver set

Protective Clothing

Goggles or safety glasses with side shields

Describe the vehicle being worked on:

Year __________ Make ____________________ Model ____________________

VIN ____________________ Engine type and size ____________________

Transmission type and model ______________________________

PROCEDURE

1. With the vehicle on a lift, check for leaks at and around the transmission's extension housing. Record your findings.

____

____

2. An oil leak stemming from the mating surfaces of the extension housing and the transmission case may be caused by loose bolts. To correct this problem, tighten the bolts to the specified torque. What is the specified torque?

____

3. Check the extension housing for cracks, especially around the case mounting surface and the pad that attaches to the transmission mount. Record your findings.

____

____

4. Also check for signs of leakage at the rear of the extension housing. Fluid leaks from the seal of the extension housing can be corrected with the transmission in the car. Record your findings.

____

____

5. If the leak was at the rear of the housing, the problem may be the bushing or seal. The clearance between the drive shaft's sliding yoke and the bushing should be minimal. If the clearance is satisfactory, a new oil seal will correct the leak. If the clearance is excessive, a new seal and a new bushing should be installed. If the seal is faulty, the transmission vent should be checked for blockage. Check the transmission vent and describe your findings.

____

____

6. To begin the procedure for replacing the bushing and seal, remove the drive shaft from the vehicle. What do you need to remember before removing the drive shaft?

____

____

7. Insert the appropriate puller tool into the extension housing until it grips the front side of the bushing. ☐ Task completed

8. Pull the seal and bushing from the housing. ☐ Task completed

9. To replace the bushing, drive a new bushing, with the appropriate driver, into the extension housing. Always make sure this bushing is aligned correctly during replacement or premature failure can result. ☐ Task completed

10. Lubricate the lip of the seal, then install the new seal in the extension housing. What did you use to install the seal?

____

11. Then, install the drive shaft. ☐ Task completed

12. If only the seal needs to be replaced, remove the old seal with a puller. ☐ Task completed

13. Lubricate the lip of the seal, then install the new seal in the extension housing. What did you use to install the seal?

____

14. Then, install the drive shaft. ☐ Task completed

15. The vehicle's speedometer can be purely electronic, which requires no mechanical hook-up to the transmission, or it can be driven off the output shaft. An oil leak at the speedometer cable can be corrected by replacing the O-ring seal. Check the seal and record the results.

____

____

16. If the seal is bad, remove the speedometer drive assembly from the extension housing by removing the hold-down screw that keeps the retainer in its bore. ☐ Task completed

17. Carefully pull up on the speedometer cable, pulling the speedometer retainer and drive gear assembly from its bore. ☐ Task completed

18. While replacing a bad seal, inspect the speedometer drive gear for chips and missing teeth. A damaged drive gear can cause the driven gear to fail, therefore both should be carefully inspected. On some transmissions the speedometer drive gear is a set of gear teeth machined into the output shaft. Inspect this gear. If the teeth are slightly rough, they can be cleaned up and smoothened with a file. If the gear is severely damaged, the entire output shaft must be replaced. Other transmissions have a drive gear that is splined to the output shaft, held in place by a clip, or driven and retained by a ball that fits into a depression in the shaft. If a clip is used, it should be carefully inspected for cracks or other damage. The drive gear can be removed and replaced, if necessary. Record your findings.

____

____

19. Carefully remove the old seal from the speedometer drive. ☐ Task completed

20. Before replacing the speedometer drive seal, clean the top of the cable's retainer. ☐ Task completed

21. Inspect the retainer of the driven gear on the speedometer cable. Record your findings.

____

____

22. Lightly grease and install the O-ring onto the retainer. Also lubricate the drive gear. ☐ Task completed

23. Gently tap the retainer and gear assembly into its bore while lining up the groove in the retainer with the screw hole in the side of the case. ☐ Task completed

24. Install the hold-down screw and tighten it in place. ☐ Task completed

25. If the cause of the leak is not found, prepare to remove the extension housing by removing the drive shaft. ☐ Task completed

26. Clean the extension housing. ☐ Task completed

27. Using a known flat surface, check the flatness of the mating surface. Any defects that cannot be removed by light filing indicate that the housing should be replaced. Record your findings.

____

____

28. Carefully inspect all bores, whether they are threaded or not. All damaged threaded areas should be repaired. If any condition exists that cannot be adequately repaired, the housing should be replaced. Record your findings.

____

____

29. Often the speedometer drive gear is responsible for throwing oil back to the rear bushing. A sheared or otherwise inoperative speedometer gear could cause the extension housing bushing to fail. What is the condition of the speedometer drive gear?

____

____

30. Before installing the rear extension housing, collect all other assemblies that may be enclosed by the extension housing. Are they are assembled properly and in good shape?

____

____

31. Install a new extension housing gasket and tighten the housing to the transmission. What are the torque specifications for those bolts?

____

32. Reinstall the drive shaft. ☐ Task completed

Instructor's Comments

____

____

____

MANUAL TRANSMISSIONS JOB SHEET 9

Road Test a Vehicle for Transmission Problems

Name ______________________________ Station ________________ Date ________________

NATEF Correlation

This Job Sheet addresses the following NATEF task:

B.4. Diagnose noise, hard shifting, jumping out of gear, and fluid leakage concerns; determine necessary action.

Objective

Upon completion of this job sheet, you will be able to demonstrate the ability to properly road test a vehicle to identify transmission problems.

Tools and Equipment

A vehicle with a manual transmission or transaxle

Protective Clothing

Goggles or safety glasses with side shields

Describe the vehicle being worked on:

Year __________ Make ________________________ Model ________________________

VIN ________________________ Engine type and size ________________________

PROCEDURE

1. While driving the vehicle in town, obey the speed laws. Before beginning the road test, check the feel of the clutch pedal. Does it feel normal? Describe the feel and action of the clutch pedal.

____

____

____

2. Shift the transmission through the gears and identify any problem you might feel.
   a. Did it easily shift into first? _____ yes ______ no
   b. Did the gear change feel smooth? ______ yes _____ no
   c. Did it easily shift into second? _____ yes ______ no
   d. Did the gear change feel smooth? ______ yes _____ no
   e. Did it easily shift into third? _____ yes ______ no
   f. Did the gear change feel smooth? ______ yes _____ no
   g. Did it easily shift into fourth? _____ yes ______ no
   h. Did the gear change feel smooth? ______ yes _____ no
   i. Did it easily shift into fifth? _____ yes ______ no
   j. Did the gear change feel smooth? ______ yes _____ no

k. Did it easily shift into reverse? _____ yes ______ no
l. Did the gear change feel smooth? ______ yes _____ no

3. Describe what you think could be the cause of any shifting problem:

____

____

4. Now drive the vehicle, in one gear at a time. Accelerate, then back off the throttle.

a. Did the transmission stay in first gear? ______ yes _____ no
b. Did the transmission stay in second gear? ______ yes _____ no
c. Did the transmission stay in third gear? ______ yes _____ no
d. Did the transmission stay in fourth gear? ______ yes _____ no
e. Did the transmission stay in fifth gear? ______ yes _____ no
f. Did the transmission stay in reverse gear? ______ yes _____ no

5. Describe what you think could be the cause of any jumping out of gear problem:

____

____

____

6. Describe the general condition of the transmission.

____

____

____

Instructor's Comments

____

____

____

MANUAL TRANSMISSIONS JOB SHEET 10

Inspect and Adjust Shift Linkage

Name ______________________ Station ______________ Date ______________

NATEF Correlation

This Job Sheet addresses the following NATEF task:

B.5. Inspect, adjust, and reinstall shift linkages, brackets, bushings, cables, pivots, and levers.

Objective

Upon completion of this job sheet, you will be able to demonstrate the ability to inspect and adjust manual transmission and transaxle shift linkages.

Tools and Materials

Droplight
Hoist
Pliers
Rags
Service manual
Shifter alignment tool
Solvent

Protective Clothing

Goggles or safety glasses with side shields

Describe the vehicle being worked on:

Year __________ Make ____________________ Model ____________________

VIN ____________________ Engine type and size ____________________

PROCEDURE (PREPARATION)

1. Place the shift lever in the neutral position. Raise the vehicle on a hoist. Wipe off all linkage parts with a rag. ☐ Task completed
2. Inspect the linkage for damage or wear. Check the shift-linkage rods for looseness and for worn or missing bushings. Record your results on the Report Sheet for Adjusting Floor-Mounted Shift Linkages. ☐ Task completed
3. Insert the shifter alignment tool to hold the linkage in its neutral position. ☐ Task completed

PROCEDURE (LINKAGE RODS WITH SLOTTED ENDS)

1. Loosen the adjustment nut with an open-end or box-end wrench until the rod is free to move. Hold the slotted end using a screwdriver inserted into the slot. Loosen the adjustment nut with an open-end or box wrench. ☐ Task completed

2. Grasp the shifting arms and move them back and forth by hand, checking for smooth and positive movement. ☐ Task completed

3. Place the shifting arms in their neutral position. When reconnecting the shift-linkage rods, adjust the rods to match the distance between the levers and arms. After the rods have been adjusted for length, tighten them in place. ☐ Task completed

4. Tighten the adjustment nut by hand until the nut begins to contact the arm. Hold the slotted end of the rod by using a screwdriver inserted into the slot. Tighten the adjustment nut with an open-end or box-end wrench. ☐ Task completed

PROCEDURE (LINKAGE RODS WITH THREADED ENDS)

1. Remove the holding pins from the swivels. Pull the swivels from the shift levers. ☐ Task completed

2. Grasp the shifting arms and move them back and forth by hand, checking for smooth and positive movement. ☐ Task completed

3. Place the shifting arms in their neutral position. When reconnecting the shift-linkage rods, the rods must be adjusted to match the distance between the levers and the arms. After the rods have been adjusted for length, tighten them in place. ☐ Task completed

4. Lubricate the bushings with white grease. Rotate the adjustment swivels until the pin ends of the swivels slip easily into the bushings in the shift levers. Install new holding pins. ☐ Task completed

   NOTE: *Always use new holding pins. Old holding pins will break if they are reused.*

PROCEDURE (LINKAGE RODS WITH CLAMPS)

1. Loosen the bolts that hold the clamps to the transmission arms. The bolt should be loosened until the rods are free to move. ☐ Task completed

2. Grasp the shifting arms and move them back and forth by hand, checking for smooth and positive movement. ☐ Task completed

3. Place the shifting arms in their neutral position. When reconnecting the shift-linkage rods, adjust them to match the distance between the levers and the arms. After the rods have been adjusted for length, tighten them in place. ☐ Task completed

4. Tighten the clamp bolts with an open-end or box-end wrench. ☐ Task completed

PROCEDURE (FINAL CHECKS—ALL TYPES)

1. Remove the alignment tool. ☐ Task completed
2. Lubricate the shift linkage with the proper lubricant. ☐ Task completed
3. Lower the vehicle. ☐ Task completed
4. Test the shift lever through all gears. The operation of the shift lever should be smooth. Note any shifting noise or hard shifting problems. Record your results on the Report Sheet for Adjusting Floor-Mounted Shift Linkages. ☐ Task completed
5. Adjust the backdrive rod. ☐ Task completed

Problems Encountered

____

____

____

Instructor's Comments

____

____

____

REPORT SHEET FOR ADJUSTING FLOOR-MOUNTED SHIFT LINKAGES		
	Serviceable	*Nonserviceable*
1. Inspection		
Linkage rods		
Bushings		
Shifting arms		
2. Final checks	*Yes*	*No*
Shift into all gears		
Smooth operation		
Jump out of gear		
Noises		
Hard shifting		
Conclusions and Recommendations ______		

MANUAL TRANSMISSIONS JOB SHEET 11

Checking the Transaxle Mounts

Name ______________________ Station ______________ Date ______________

NATEF Correlation

This Job Sheet addresses the following NATEF task:

B.6. Inspect and reinstall power train mounts.

Objective

Upon completion of this job sheet, you will be able to inspect, replace, and align power train mounts.

Tools and Materials

Engine support fixture
Engine hoist
Basic hand tools

Protective Clothing

Goggles or safety glasses with side shields

Describe the vehicle being worked on:

Year ________ Make ____________________ Model ____________________

VIN ____________________ Engine type and size ____________________

Transmission type and model ____

PROCEDURE

1. Many shifting and vibration problems can be caused by worn, loose, or broken engine and transmission mounts. Visually inspect the mounts for looseness and cracks. Give a summary of your visual inspection.

____

____

____

2. Pull up and push down on the transaxle case while watching the mount. If the mount's rubber separates from the metal plate or if the case moves up but not down, replace the mount. If there is movement between the metal plate and its attaching point on the frame, tighten the attaching bolts to an appropriate torque. Describe the results of doing this.

____

____

3. From the driver's seat, apply the foot brake, set the parking brake, and start the engine. Put the transmission in first gear, raise the engine speed to about 1500–2000 rpm, and gradually release the clutch pedal. Watch the torque reaction of the engine on its mounts. If the engine's reaction to the torque appears to be excessive, broken or worn drive train mounts may be the cause. Describe the results of doing this.

____

____

4. If it is necessary to replace the transaxle mount, make sure you follow the manufacturer's recommendations for maintaining the alignment of the driveline. Describe the recommended alignment procedures.

____

____

____

____

5. When removing the transaxle mount, begin by disconnecting the battery's negative cable. ☐ Task completed

6. Disconnect any electrical connectors that may be located around the mount. Be sure to label any wires you remove to facilitate reassembly. ☐ Task completed

7. It may be necessary to move some accessories, such as the horn, in order to service the mount without damaging some other assembly. ☐ Task completed

8. Install the engine support fixture and attach it to an engine hoist. ☐ Task completed

9. Lift the engine just enough to take the pressure off the mounts. ☐ Task completed

10. Remove the bolts attaching the transaxle mount to the frame and the mounting bracket, then remove the mount. ☐ Task completed

11. To install the new mount, position the transaxle mount in its correct location on the frame and tighten its attaching bolts to the proper torque. What is the torque specification?

____

12. Install the bolts that attach the mount to the transaxle bracket. Prior to tightening these bolts, check the alignment of the mount. ☐ Task completed

13. Once you have confirmed that the alignment is correct, tighten all loosened bolts to their specified torque. ☐ Task completed

14. Remove the engine hoist fixture from the engine and reinstall all accessories and wires that may have been removed earlier. ☐ Task completed

Instructor's Comments

____

____

____

MANUAL TRANSMISSIONS JOB SHEET 12

Sealing a Transmission or Transaxle

Name ______________________ Station ______________ Date ______________

NATEF Correlation

This Job Sheet addresses the following NATEF task:

B.7. Inspect and replace gaskets, seals, and sealants; inspect sealing surfaces.

Objective

Upon completion of this job sheet, you will be able to inspect and replace gaskets, seals, and sealants, and inspect sealing surfaces.

Tools and Materials

Basic hand tools
Thread repair kit
Fine file

Protective Clothing

Goggles or safety glasses with side shields

Describe the vehicle being worked on:

Year ________ Make __________________ Model ____________________

VIN ____________________ Engine type and size ____________________

Transmission type and number of forward speeds ______________________

PROCEDURE

1. Inspect the transmission and clutch for cracks, worn or damaged bearing bores, damaged threads, or any other damage that could affect the operation of the transmission/transaxle. Describe the results of your inspection.

____

____

2. Inspect the transmission case and clutch housing mating surfaces for small nicks or burrs that could cause an oil leak or a misalignment of the two halves. These nicks and burrs can be removed with a fine stone or file. Describe the results of your inspection.

____

____

3. Check the extension housing for cracks and repair or replace it as needed. Describe the results of your inspection.

____

____

4. Check the mating surfaces of the housing for burrs or gouges and file the surface flat. Describe the results of your inspection.

____

____

5. Check all threaded holes and repair any damaged bores with a thread repair kit. Describe the results of your inspection.

____

____

6. Check the bushing in the rear of the extension housing for excessive wear or damage. Always replace the rear extension seal. Describe the results of your inspection.

____

____

7. Remove the bearing from the extension and check it for smooth and quiet rotation. If the bearing is dry, it may sound noisy; if a few drops of clean oil are applied to a good bearing, the noise will cease. Check all bearing cups for excessive wear or damage. Describe the results of your inspection.

____

____

8. Inspect the center support plate (if the trans has one) for cracks. Describe the results of your inspection.

____

____

9. Carefully inspect the counter gear bearing for pitting or any other signs of failure. Describe the results of your inspection.

____

____

10. Inspect the front bearing retainer for cracks, burrs, or gouges on the case mating surface. Describe the results of your inspection.

____

____

11. Inspect the retainer's snout. It should be smooth. Describe the results of your inspection.

____

____

12. Remove any gasket material from the sealing surface and install a new shaft seal before reinstalling the bearing retainer. Make sure the oil return hole in the retainer and the gasket lines up with the oil hole in the case. ☐ Task completed

Instructor's Comments

____

____

____

MANUAL TRANSMISSIONS JOB SHEET 13

Servicing FWD Final Drives

Name ______________________ Station ______________ Date ______________

NATEF Correlation

This Job Sheet addresses the following NATEF tasks:

B.8. Remove and replace transaxle final drive.

B.14. Remove, inspect, measure, adjust, and reinstall transaxle final drive, pinion gears (spiders), shaft, side gears, side bearings, thrust washers, and case assembly.

Objective

Upon completion of this job sheet, you will be able to remove, inspect, measure, adjust, and reinstall the final drive assembly in a transaxle.

Tools and Materials

Service manual
Dial indicator and mounting fixtures
Inch-pound torque wrench
Feeler gauge
0 to 1 inch micrometer
Set of gauging shims
Fresh lubricant

Protective Clothing

Goggles or safety glasses with side shields

Describe the vehicle being worked on:

Year __________ Make ______________________ Model ______________________

VIN ______________________ Engine type and size ______________________

Transmission type and number of forward speeds ______________________

PROCEDURE

1. With the final drive assembly out of the transaxle case, locate the procedure for disassembly and assembly of the unit in the service manual. The following procedure will apply to most units. If you need to do something different, write down those differences in the appropriate space below. What service manual are you using?

   ____

2. Separate the ring gear from the differential case. The ring gear of many transaxles is riveted to the differential case. The rivets must be drilled, then driven out with a hammer and drift to separate the ring gear from the case. ☐ Task completed

3. Remove the pinion shaft lock bolt. ☐ Task completed

4. Remove the pinion shaft, then remove the gears and thrust washers from the case. ☐ Task completed

5. The side bearings of most final drive units must be pulled off and pressed onto the differential case. Always be sure to use the correct tools for removing and installing the bearings. ☐ Task completed

6. While you are disassembling the differential or transaxle, make sure to keep all shims and bearing races together and identified so that they can be reinstalled in their original location. ☐ Task completed

7. Carefully inspect the bearings for wear and/or damage and determine whether a bearing should be replaced. What were your findings?

   ____

   ____

8. The final drives of most transaxle differential cases are fitted with a speedometer gear pressed onto the case and under a side bearing. These gears are pulled off and pressed on the case. ☐ Task completed

9. Clean and inspect all parts. Replace parts as required. What parts need to be replaced?

   ____

   ____

   ____

10. When it is necessary to replace a bearing, race, or housing, refer to the manufacturer's recommendation for nominal shim thickness. If only other parts of the differential or transaxle are replaced, reuse the original shims. When repairs require the use of a service shim, discard the original shim. Never use the original shim together with the service shim. The shims must be installed only under the bearing cups at the transaxle case end of both the input and output shafts. ☐ Task completed

11. Install the gears and thrust washers into the case and install the pinion shaft and lock bolt. Tighten the bolt to the specified torque. What is that torque?

    ____

12. Attach the ring gear to the differential case. If the ring gear was riveted to the case, use the proper replacement nuts and bolts. These nuts and bolts must be of the specified hardness and should be tightened in steps and to the specified torque. What is that torque?

    ____

13. Selection of the preload shims for reassembly can begin when the input and output shaft assemblies and the differential assembly are reassembled and are ready to be installed into the transaxle case. Always check your shop manual for the exact procedure to follow. ☐ Task completed

Side Gear Endplay

This procedure is typical for measuring and adjusting the side gear endplay in a transaxle's final drive. Always refer to your service manual before proceeding to make these adjustments on a transaxle.

1. Install the correct adapter into the differential bearings. ☐ Task completed
2. Mount dial indicator to the ring gear with the plunger resting against the adapter. ☐ Task completed
3. With your fingers or a screwdriver, move the ring gear up and down. ☐ Task completed
4. Record the measured endplay.

   ____

5. Measure the old thrust washer with the micrometer. What was your reading?

   ____

6. Install the correct size shim. ☐ Task completed
7. Repeat the procedure for the other side. ☐ Task completed

Bearing Preload

The following procedure is typical for the measurement and adjustment of the differential bearing preload in a transaxle. Always refer to your service manual before proceeding to make these adjustments on a transaxle.

1. Remove the bearing cup and existing shim from the differential bearing retainer. ☐ Task completed
2. What size shim will allow for 0.001 to 0.010 inch endplay?

   ____

3. Install the gauging shim into the differential bearing retainer. ☐ Task completed
4. Press in the bearing cup. ☐ Task completed
5. Lubricate the bearings and install them into the case. ☐ Task completed
6. Install the bearing retainer. ☐ Task completed
7. Tighten the retaining bolts. ☐ Task completed
8. Mount the dial indicator with its plunger touching the differential case. ☐ Task completed
9. Apply medium pressure in a downward direction while rolling the differential assembly back and forth several times. ☐ Task completed
10. Zero the dial indicator. ☐ Task completed
11. Apply medium pressure in an upward direction while rotating the differential assembly back and forth several times. What reading did you have?

    ____

12. The required shim to set preload is the thickness of the gauging shim plus the recorded endplay. ☐ Task completed

13. Remove the bearing retainer, cup, and gauging shim. ☐ Task completed

14. Install the required shim. ☐ Task completed

15. Press the bearing cup into the bearing retainer. ☐ Task completed

16. Install the bearing retainer and tighten the bolts. ☐ Task completed

17. Check the rotating torque of the transaxle. If this is less than specifications, install a thicker shim. If the torque is too great, install a slightly thinner shim. What were the specifications and what did you need to do in order to meet that specification?

    ____

    ____

18. Repeat the procedure until desired torque is reached. ☐ Task completed

Instructor's Comments

____

____

____

MANUAL TRANSMISSIONS JOB SHEET 14

Inspecting Internal Shift Mechanisms

Name ______________________ Station ______________ Date ______________

NATEF Correlation

This Job Sheet addresses the following NATEF task:

B.9. Inspect, adjust, and reinstall shift cover, forks, levers, grommets, shafts, sleeves, detent mechanism, interlocks, and springs.

Objective

Upon completion of this job sheet, you will be able to inspect, adjust, and reinstall shift mechanisms.

Tools and Materials

Straightedge

Protective Clothing

Goggles or safety glasses with side shields

Describe the vehicle being worked on:

Year ________ Make ____________________ Model ____________________

VIN ____________________ Engine type and size ____________________

Transmission type and number of forward speeds ________________________

PROCEDURE

1. The shift forks should be inspected to make sure they are not bent, cracked, or broken. Describe the results of your inspection.

   ____

   ____

2. Slide the forks over the shift rail; they should slide easily but without much play or wobble. Describe the results of your inspection.

   ____

   ____

3. Inspect the shift rails to be sure they are not worn, bent, cracked, or broken. Describe the results of your inspection.

   ____

   ____

4. Place the rails into their bores in the case and/or extension housing and check the rails' wear by noting the movement of the rail in the bores. Describe the results of your inspection.

____

____

5. Using a straightedge, check the interlock plates for flatness. The plates should be flat and free of any evidence of excessive wear, especially on the surface that rides against the shift rails. Describe the results of your inspection.

____

____

6. With the straightedge, determine if the detent is out of alignment with the interlock. The straightedge should rest on both sides of the interlock without interference from the detent. If there is some interference on either side, place the detent and interlock assembly in a vise and apply light pressure on the detent to push it into alignment. Then loosen the nut securing the detent spring to the interlock. The spring is slotted beneath the nut and it will align itself when the tension is relieved. Now tighten the detent spring to the interlock nut. Recheck the detent alignment with a straightedge. Describe the results of your inspection.

____

____

Instructor's Comments

____

____

____

MANUAL TRANSMISSIONS JOB SHEET 15

Servicing Gears and Synchronizers

Name ______________________ Station ______________ Date ______________

NATEF Correlation

This Job Sheet addresses the following NATEF task:

B.11. Inspect and reinstall synchronizer hub, sleeve, keys (inserts), springs, and blocking rings.

Objective

Upon completion of this job sheet, you will be able to inspect gears and inspect and reinstall a synchronizer hub, sleeve, keys, springs, and blocking rings.

Tools and Materials

Basic hand tools

Feeler gauge set

Protective Clothing

Goggles or safety glasses with side shields

Describe the vehicle being worked on:

Year ________ Make ____________________ Model ____________________

VIN ____________________ Engine type and size ____________________

Transmission type and number of forward speeds ________________________

PROCEDURE

1. Carefully examine the gears of the transmission and describe the wear patterns on the teeth. All gears should show wear patterns in the center of their teeth. These wear patterns should appear as a polished finish with little wear on the gear face. Describe the results of your inspection.

   ____

   ____

2. Check the gears' teeth carefully for chips, pitting, cracks, and breakage. Describe the results of your inspection.

   ____

   ____

3. Inspect the locking teeth on the gears. Describe the results of your inspection.

   ____

   ____

4. Inspect the gears' center bores. Describe the results of your inspection.

___

___

5. Check the cone-shaped area of each gear for nicks, burrs, or gouges. Describe the results of your inspection.

___

___

6. Inspect all needle bearings; they should be smooth and shiny. Describe the results of your inspection.

___

___

7. Carefully inspect the input shaft. Check the tip that rides in the pilot bearing for smoothness. Also check the splines of the shaft for any wear that might prevent the disc from sliding evenly and smoothly. Describe the results of your inspection.

___

___

8. Inspect the area of the shaft that the oil seal rides on; this area should be free of all imperfections, because a scratch or nick could cause the oil seal to leak. Rotate the bearing; it should rotate smoothly and quietly. Describe the results of your inspection.

___

___

9. Check the gear for cracks, pitting, and chipped or broken teeth. It is normal for the front of gear teeth to show some signs of wear. The dogteeth should be pointed and not cracked or broken. Describe the results of your inspection.

___

___

10. Remove the roller bearings from the shaft and inspect the bore for burrs or pits. Check the bearing's rollers for roughness, pitting, or any other signs of failure. All bearings should be examined for signs of wear or overheating. Describe the results of your inspection.

___

___

11. Reverse gear should be inspected like any other gear, except for sliding reverse gears. These gears do not have all the features of forward gears, so inspection is limited to the gear teeth and the center bore and bearings. Describe the results of your inspection.

___

___

12. Inspect the reverse idler gear for pitted, cracked, nicked, or broken teeth. Check its center bore for a smooth and mar-free surface. Describe the results of your inspection.

____

____

13. Carefully inspect the needle bearing that the idler gear rides on for wear, burrs, and other defects. Also, inspect the reverse idler gear shaft's surface for scoring, wear, and other imperfections. Describe the results of your inspection.

____

____

14. Inspect the bearing surfaces of the main shaft; they should be smooth and show no signs of overheating. Describe the results of your inspection.

____

____

15. Inspect the gear journal areas on the shaft for roughness, scoring, and other defects. Describe the results of your inspection.

____

____

16. Check the shaft's splines for wear, burrs, and other conditions that would interfere with the slip yoke being able to slide smoothly on the splines. Describe the results of your inspection.

____

____

17. Check the main shaft bearing for smooth and quiet rotation. Describe the results of your inspection.

____

____

18. Gather the synchronizer assemblies and check each for smooth operation. Describe the results of your inspection.

____

____

19. Disassemble the units and inspect each part as follows:

    a. Carefully inspect the assemblies for worn teeth and damaged cone surfaces. The dogteeth of the synchros may appear to be slightly rounded; this is normal and will not interfere with normal operation. Describe the results of your inspection.

    ____

    ____

b. Move the synchronizer sleeves on their hubs and feel for any imperfections that may suggest the sleeve is not able to move freely on the shaft. The sleeve's inner splines should be pointed. The sleeve should also be carefully inspected for signs of fatigue, such as cracks or chips. Describe the results of your inspection.

____

____

c. Make sure the alignment marks on the sleeve and hub are properly positioned.

d. Examine the position of the insert springs. Also the springs should not be bent, distorted, or broken. Describe the results of your inspection.

____

____

e. Inspect the blocking rings for wear marks on the face of the splined end that may indicate that the ring was bottoming on the gear face due to wear of the blocker ring. The blocking ring can also be checked for wear by placing a feeler gauge between the ring and the gear's dogteeth. Compare your measurement to the manufacturer's specifications. Describe the results of your inspection.

____

____

f. The rings should also be inspected for cracks, breakage, and flatness. To check the flatness of the ring, place it on a flat surface and check how flatly it sits on the surface. The dogteeth should be pointed and have smooth surfaces for the inner splines of the synchronizer sleeve to ride on. Check the threaded cone area; the threads should be well defined and sharp. Describe the results of your inspection.

____

____

g. The hub should be inspected to make sure that it is not bent or cracked or has signs of possible failure. Describe the results of your inspection.

____

____

Instructor's Comments

____

____

____

MANUAL TRANSMISSIONS JOB SHEET 16

Road Check Differential Noises

Name ______________________ Station ______________ Date ______________

NATEF Correlation

This Job Sheet addresses the following NATEF tasks:

B.13. Diagnose transaxle final drive assembly noise and vibration concerns; determine necessary action.

D.1.1. Diagnose noise and vibration concerns; determine necessary action.

Objective

Upon completion of this job sheet, you will be able to road check a vehicle and identify noises and vibration concerns in the differential.

Tools and Equipment

A RWD vehicle

Describe the vehicle being worked on:

Year ________ Make __________________ Model __________________

VIN __________________ Engine type and size __________________

Describe general operating condition:

____

PROCEDURE

1. Check the fluid level in the rear axle. Check the vent cap on the housing to make sure it is open. ☐ Task completed

2. Road test the vehicle. It should be driven for enough time to warm it up. Drive about 55 mph; move the throttle pedal so that the vehicle's speed increases and decreases. Drive at one speed for a short while; do not accelerate hard. Describe any abnormal noises and note when they occur:

____

____

____

3. Now accelerate quickly, then let off the throttle. Describe any abnormal noises:

____

____

____

4. Based on the above checks, what are your conclusions about the final drive assembly?

____

____

____

Instructor's Comments

____

____

____

MANUAL TRANSMISSIONS JOB SHEET 17

Check Fluid in a Manual Transmission and Transaxle

Name ______________________ Station ______________ Date ______________

NATEF Correlation

This Job Sheet addresses the following NATEF task:

B.15. Inspect lubrication devices (oil pump or slingers); perform necessary action.

Objective

Upon completion of this job sheet, you will be able to demonstrate the ability to check the fluid level in a manual transmission and transaxle.

Tools and Materials

Hand tools

Service manual

Protective Clothing

Goggles or safety glasses with side shields

Describe the vehicle being worked on:

Year ________ Make ____________________ Model ____________________

VIN ____________________ Engine type and size ____________________

PROCEDURE

1. Refer to the service manual to determine the fluid level check point on the specific vehicle you are checking. ☐ Task completed
2. Refer to the service manual to determine the type of fluid for the specific vehicle you are checking. ☐ Task completed
3. The transmission/transaxle gear oil level should be checked at the intervals specified in the service manual. Normally, these range from every 7,500 to 30,000 miles. For service convenience, many units are now designed with a dipstick and filler tube accessible from beneath the hood. Check the oil with the engine off and the vehicle resting on level grade. If the engine has been running, wait 2 to 3 minutes before checking the gear oil level. ☐ Task completed
4. Some vehicles have no dipstick. Instead, the vehicle must be placed on a lift and the oil level checked through the fill plug opening on the side of the unit. Clean the area around the plug before loosening and removing it. Insert a finger or bent rod into the hole to check the level. The oil may be hot. Lubricant should be level with, or not more than 1/2 inch below the fill hole. Add the proper grade lubricant as needed using a filler pump. ☐ Task completed

5. Manual transmission/transaxle lubricants in use today include single and multiple viscosity gear oils, engine oils, synchromesh fluid and automatic transmission fluid. Always refer to the service manual to determine the correct lubricant and viscosity range for the vehicle and operation conditions.

☐ Task completed

Problems Encountered

____

____

____

Instructor's Comments

____

____

____

MANUAL TRANSMISSIONS JOB SHEET 18

Testing Transmission Sensors and Switches

Name ____________________ Station ______________ Date ______________

NATEF Correlation

This Job Sheet addresses the following NATEF task:

B.16. Inspect, test, and replace transmission or transaxle sensors and switches.

Objective

Upon completion of this job sheet, you will be able to inspect and test, adjust, repair, or replace transmission sensors and switches.

Tools and Materials

Measuring tape

DMM

Jack and safety stands or hoist to raise the vehicle

12-volt test light

Component locator

Service manual

Protective Clothing

Goggles or safety glasses with side shields

Describe the vehicle being worked on:

Year __________ Make ____________________ Model ____________________

VIN ____________________ Engine type and size ____________________

PROCEDURE

1. Using the service manual and component locator for the vehicle, list all of the switches and sensors located on the transmission or transaxle.

   ____

   ____

   ____

2. Carefully check all electrical wires and connectors for damage, looseness, and corrosion. Record your findings.

   ____

   ____

3. Use an ohmmeter to check the continuity through a connector suspected of being faulty. Record your findings.

____

____

4. Check all ground cables and connections. Corroded battery terminals and/or broken or loose ground straps to the frame or engine block will cause problems. Record your findings.

____

____

5. Check the fuse or fuses to the related circuits. To accurately check a fuse, either test it for continuity with an ohmmeter or check each side of the fuse for power when the circuit is activated. Record your findings.

____

____

6. With the system on, measure the voltage dropped across connectors and circuits. Record your findings.

____

____

7. Switches can be tested for operation and for excessive resistance with a voltmeter, test light, or ohmmeter. To check the operation of a switch with a voltmeter or a test light, connect the meter's positive lead to the battery side of the switch. With the negative lead attached to a good ground, voltage should be measured at this point.

☐ Task completed

8. Without closing the switch, move the positive lead to the other side of the switch. If the switch is open, no voltage will be present at that point. The amount of voltage present at this side of the switch should equal the amount on the other side when the switch is closed. If the voltage decreases, the switch is causing a voltage drop due to excessive resistance. If no voltage is present on the groundside of the switch with it closed, the switch is not functioning properly and should be replaced. Record your findings.

____

____

9. If a switch has been removed from the circuit, it can be tested with an ohmmeter or a self-powered test light. By connecting the leads across the switch connections, the action of the switch should open and close the circuit. Record your findings.

____

____

10. The pressure switches used in today's transmissions are either grounding switches or they connect or disconnect two wires. Refer to the wiring diagram to determine the type of switch and test the switch with an ohmmeter. Base your expected results on the type of switch you are testing. By using air pressure, you can easily see if the switch works properly or if it has a leak. Record your findings.

____

____

11. Wheel speed sensors provide road speed information to the computer. There are basically two types of speed sensors, AC voltage generator sensors and reed style sensors. Both of these rely on magnetic principles. The activity of the sensor can be checked with an ohmmeter. AC voltage generators rely on a stationary magnet and a rotating shaft fitted with iron teeth. Each time a tooth passes through the magnetic field, an electrical pulse is present. By counting the number of teeth on the output shaft, you can determine how many pulses per revolution you will measure with a voltmeter set to AC volts. How many teeth are on the output shaft? How many pulses on the meter did you observe?

____

____

____

Clutch Pedal Switch

1. Find the specifications for the following: (If exact specifications are not available, use the general specifications given in the text.)

   Pedal free play ____________________

   Clutch pedal height ____________________

   Pedal travel ____________________

   Clearance between clutch switch and pedal assembly ________________

2. Measure the following and compare to specifications:

   Pedal free play ____________________

   Clutch pedal height ____________________

3. If the measurements from step 2 are not within specifications, correct them before proceeding.

4. Check the release point of the clutch.
   a. Engage the parking brake and install wheel chocks.
   b. Start the engine and allow it to idle.
   c. Without depressing the clutch pedal, slowly move the shift lever into reverse until the gears contact.
   d. Gradually depress the clutch pedal until the noise stops. This is the release point.

5. Measure the stroke distance from the release point up to the full stroke end position.

   Your measurement ____________________

   If this distance is not within specifications, check the pedal height, push rod play, and pedal freeplay. Adjust the pedal for the correct travel.

6. If the travel is correct, check the clearance between the switch and the pedal assembly when the clutch is fully depressed.

   Your measurement ____________________

7. If the clearance is not within specifications, loosen the switch and adjust its position to provide for the specified clearance.

8. Tighten switch to specifications.

9. Work the clutch pedal and note any noises. If there are no unusual noises, the task is complete. If there are noises, find the cause and correct the problem.

Reverse Light Switch

1. Check the fuse for the backup lights.

   Condition ____________________

2. Move the gear shift slightly in all directions with the ignition switch ON but the engine OFF.

   Do the backup lights come on now? ____________________

3. If the backup lights come on or flicker when the gear shift is moved, the backup light switch probably needs to be adjusted.

   Sometimes a loose or worn shift linkage will cause the same problem. Check the linkage for tightness. Condition of linkage ____________________

4. If the backup lights still do not operate, check the condition of the wires and connectors at the backup lamp assemblies.

   Condition ____________________

5. Check the condition of the lamps. If the condition of a bulb is questionable, check it with an ohmmeter.

   Condition of the bulbs ____________________

6. Check the condition of the wires and connectors at the backup light switch.

   Condition ____________________

7. Correct any problem found and recheck the lights. If the backup lights still do not work, proceed.

8. Check the switch by measuring the resistance across the switch's terminals.

   Resistance reading with shift lever in reverse ____________________

   Resistance reading with shift lever in a forward gear ____________________

   There should be good continuity only when the gear shift lever is in the reverse position. If the switch remains open when the transmission is in reverse, the switch needs to be replaced.

9. If there is continuity across the switch's terminals when the shift lever is in reverse and the backup lights don't work, check the circuit for an open.

Instructor's Comments

____

____

____

MANUAL TRANSMISSIONS JOB SHEET 19

Drive Axle Inspection and Diagnosis

Name ______________________ Station ______________ Date ______________

NATEF Correlation

This Job Sheet addresses the following NATEF task:

C.1. Diagnose constant-velocity (CV) joint noise and vibration concerns; determine necessary action.

Objective

Upon completion of this job sheet, you will be able to inspect and diagnose the front axles and joints on a FWD vehicle.

Tools and Equipment

A FWD vehicle

Protective Clothing

Goggles or safety glasses with side shields

Describe the vehicle being worked on:

Year ________ Make ____________________ Model ____________________

VIN ____________________ Engine type and size ____________________

Describe general condition:

____

CAUTION: *While driving the vehicle, make sure you obey all laws and have the permission of the owner to drive the vehicle.*

PROCEDURE

1. Raise the vehicle on a hoist. ☐ Task completed
2. Carefully look at the CV joint boots on both drive axles and describe their condition and your recommendation:

   ____

   ____

   ____

3. Check the tightness of the boot clamps. Record your findings:

   ____

   ____

   ____

4. Visually inspect the shafts and describe their condition:

___

5. If problems were found during the road test, describe what service needs to be performed.

___

___

6. If there are no obvious problems with the boots, clamps, and shafts, lower the vehicle and prepare it for a road test. ☐ Task completed

7. Do a quick safety check of the vehicle; this should include the tires and lights. ☐ Task completed

8. Drive the vehicle straight on the road. Pay attention to any unusual noises or handling problems. Describe your findings:

___

___

___

9. While moving straight at a low speed, accelerate hard, then let off the throttle. Pay attention to any unusual noises or handling problems. Describe your findings:

___

___

___

10. Now turn the vehicle to the right. Pay attention to any unusual noises or handling problems. Describe your findings:

___

___

___

11. Now turn the vehicle to the left. Pay attention to any unusual noises or handling problems. Describe your findings:

___

___

12. Return to the shop and record your conclusions from the road test.

___

___

___

Instructor's Comments

___

___

___

MANUAL TRANSMISSIONS JOB SHEET 20

Checking U-Joints and the Driveshaft

Name ______________________ Station ______________ Date ______________

NATEF Correlation

This Job Sheet addresses the following NATEF tasks:

C.2. Diagnose universal joint noise and vibration concerns; perform necessary action.

C.6. Check shaft balance; measure shaft runout; measure and adjust driveline angles.

Objective

Upon completion of this job sheet, you will be able to inspect the components of a RWD driveline, check and correct U-joint angles and driveshaft runout, and balance a driveshaft.

Tools and Equipment

Brass drift	Dial indicator
Miscellaneous hand tools	Torque wrench
Rags	Transmission jack
Chalk, crayon, or paint stick	Inclinometer
Hose clamps	Service manual

Protective Clothing

Goggles or safety glasses with side shields

Describe the vehicle being worked on:

Year ________ Make ____________________ Model ____________________

VIN ____________________ Engine type and size ____________________

Describe general condition:

____

PROCEDURE (VISUAL INSPECTION)

1. Place the transmission in neutral and raise the vehicle on a drive-on hoist. Check for leaks at the slip joint, U-joints, final drive pinion seal, and pinion companion flange. Record the results on the Report Sheet for Driveline Inspection. ☐ Task completed

2. Shake and twist the driveshaft to locate worn or loose parts. Pry with a screwdriver around the U-joints. Record the results on the Report Sheet for Driveline Inspection. ☐ Task completed

3. Check for dirt, undercoating, dents, or missing balancing weights on the driveshaft. Inspect the center-bearing rubber bushing and support bracket, if equipped. Record the results on the Report Sheet for Driveline Inspection. ☐ Task completed

CAUTION: *Before attempting to check a center bearing, be sure the driving wheels and driveshaft are free to rotate.*

4. Check the center bearing, if equipped. ☐ Task completed

WARNING: *Extreme care should be taken when working around a rotating driveshaft. Severe injury can result from touching a moving shaft.*

PROCEDURE (CHECK U-JOINT ANGLES)

1. Locate the specification for U-joint angles in the service manual. Clean the surfaces where the inclinometer will be mounted. ☐ Task completed

CAUTION: *Do not force the inclinometer when setting it into position, or a false reading will be recorded.*

2. Check the front U-joint angle, and record the reading on the Report Sheet for Driveline Inspection. ☐ Task completed

3. Check the rear U-joint angle, and record the reading on the Report Sheet for Driveline Inspection. ☐ Task completed

4. If necessary, correct the U-joint angles. ☐ Task completed

CAUTION: *Do not use too many shims. Measure at the center of each shim. It should be no thicker than 1/4 inch (6.35 mm). If the rear U-joint angle is not correct, other problems may exist in the suspension. These problems include broken springs or an improperly placed spring seat.*

PROCEDURE (CHECK DRIVESHAFT RUNOUT)

1. Locate the specification for driveshaft runout in the service manual. Clean the areas on the driveshaft where the dial indicator plunger will ride. ☐ Task completed

2. Mount the dial indicator. Take runout readings at each end and at the center of the driveshaft. ☐ Task completed

3. If necessary, disconnect the driveshaft, rotate it 180 degrees on the differential companion flange, and reinstall it. Recheck the runout readings. ☐ Task completed

4. If necessary, replace the driveshaft. Recheck the runout readings. ☐ Task completed

PROCEDURE (BALANCE DRIVESHAFT)

1. Raise the vehicle on a hoist and support the rear axle housings. ☐ Task completed

2. Remove the rear wheels. Replace and tighten the lug nuts with the flat edge against the brake drums. ☐ Task completed

WARNING: *Do not drive the rear axle without first locking the drums to the axles with the lug nuts. Failure to do so may result in a drum flying off of the axle, causing severe injury.*

3. Clean the driveshaft thoroughly and determine the heavy spot on the driveshaft. ☐ Task completed

WARNING: *Keep the chalk or crayon away from the balancing weights on the driveshaft. Touching the balancing weights could cause serious injury.*

4. Add two hose clamps so that the screw part is on the light side of the driveshaft. Recheck for vibration and noise. ☐ Task completed
5. If necessary, move the hose clamps around the driveshaft to balance it. ☐ Task completed
6. Recheck for vibration and noise. ☐ Task completed
7. If necessary, disconnect the driveshaft, rotate it 180 degrees on its differential companion flange, and reinstall it. ☐ Task completed
8. If necessary, move the hose clamps to the front of the driveshaft. Recheck for vibration and noise. ☐ Task completed
9. If necessary, replace the driveshaft. Recheck for vibration and noise. ☐ Task completed
10. Road-test the vehicle and make a note of any vibration and noise. ☐ Task completed

Problems Encountered

___

___

___

Instructor's Comments

___

___

___

Name ______________________ Station ______________ Date ______________

REPORT SHEET FOR DRIVELINE INSPECTION		
	Serviceable	*Nonserviceable*
1. Inspection		
Leaks		
U-joints		
Yokes/Companion flange		
Driveshaft		
2. U-joint angles		
Front angle specification		
Actual front angle		
Rear angle specification		
Actual rear angle		
3. Driveshaft runout		
Maximum runout specification		
Actual runout at front		
Actual runout at middle		
Actual runout at rear		
	Yes	*No*
4. Balance driveshaft		
Vibration		
Vibration with hose clamps installed		
Vibration with clamps moved		
Vibration with clamps at front		
Vibration with shaft rotated 180 degrees		
Vibration with new shaft		

Conclusions and Recommendations ______________________________________

____

____

MANUAL TRANSMISSIONS JOB SHEET 21

Servicing FWD Wheel Bearings

Name ______________________ Station ________________ Date ________________

NATEF Correlation

This Job Sheet addresses the following NATEF task:

C.3. Replace front-wheel-drive (FWD) front wheel bearing.

Objective

Upon completion of this job sheet, you will be able to replace front-wheel-drive front wheel bearings.

Tools and Materials

Lift
Torque wrench
Hydraulic press
Various drivers and pullers
Brass drift
Soft-faced hammer
CV-joint boot protector

Protective Clothing

Goggles or safety glasses with side shields

Describe the vehicle being worked on:

Year __________ Make ______________________ Model ______________________

VIN ______________________ Engine type and size ______________________

PROCEDURE

NOTE: *This is a typical procedure. Check with the service manual for the exact procedure you should follow and mark the differences in procedure as you progress through the steps below.*

1. Loosen the hub nut and wheel lug nuts. Have someone apply the brakes with the vehicle's weight on its tires. Loosen the axle hub nut and wheel lug nuts with the appropriate tools. Then, remove the axle hub nut. ☐ Task completed

2. Jack up the vehicle and remove the tire and wheel assembly. At this point it's a good idea to install a boot protector over the CV joint boot to protect it while you are working. ☐ Task completed

3. Unbolt the front brake caliper, move it out of the way, and suspend it with wire. Make sure the flexible brake hose doesn't support the caliper's weight. ☐ Task completed

4. Remove the brake rotor. ☐ Task completed

5. Then, loosen and remove the pinch bolt that holds the lower control arm to the steering knuckle and separate the lower ball joint and tie rod end from the knuckle. ☐ Task completed

6. On some cars, an eccentric washer is located behind one or both of the bolts that connect the spindle to the strut. These eccentric washers are used to adjust camber. Mark the exact location and position of this washer on the strut before you remove it. This allows you to reinstall the washer and the assembly without drastically disturbing the car's camber angle. ☐ Task completed
7. Once the washer is marked, remove the hub-and-bearing to knuckle retaining bolts. ☐ Task completed
8. Remove the cotter pin and castellated nut from the tie rod end and pull the tie rod end from the knuckle. ☐ Task completed
9. The tie rod end can usually be freed from its bore in the knuckle by using a small gear puller or by tapping on the metal surrounding the bore in the knuckle. ☐ Task completed
10. Remove the pinch bolts holding the steering knuckle to the strut, and then remove the knuckle from the car. ☐ Task completed
11. With a soft-faced hammer, tap on the hub assembly to free it from the knuckle. The brake backing plate may come off with the hub assembly. ☐ Task completed
12. Install the appropriate puller and press the hub-and-bearing assembly from the halfshaft. ☐ Task completed
13. Pull the halfshaft away from the steering knuckle. Do not allow the CV joint to drop downward while you remove the hub assembly. Support the halfshaft and the CV joint from the back side of the knuckle. ☐ Task completed
14. With the steering knuckle off, remove the snap ring or collar that retains the wheel bearing. ☐ Task completed
15. Mount the knuckle on a hydraulic press so that it rests on the base of the press with the hub free. Using a driver slightly smaller than the inside diameter of the bearing, press the bearing out of the hub. On some cars, this will cause half of the inner race to break out. Since half of the inner race is still on the hub, it must be removed. To do this, use a small gear puller. ☐ Task completed
16. Inspect the bearing's bore in the knuckle for burrs, score marks, cracks, and other damage. ☐ Task completed
17. Lightly lubricate the outer surface of the new bearing assembly and the inner bore of the knuckle. ☐ Task completed
18. Press the new bearing into the knuckle. Make sure all inner snap rings are in place. ☐ Task completed
19. Install the outer retaining ring and bolt the brake splash shield onto the knuckle. ☐ Task completed
20. Using the proper tool, press the new steering knuckle seal into place. Lubricate the lip seal and coat the inside of the knuckle with a thin coat of grease. ☐ Task completed
21. Slide the knuckle assembly over the drive axle. ☐ Task completed
22. Mount the knuckle to the strut and tighten the bolts. Make sure the camber marks to the strut are positioned properly. ☐ Task completed

23. Reinstall the lower control arm ball joint and tie rod end to the steering knuckle. Then mount the rotor and brake caliper. Torque all bolts to specifications. What are the specifications?

____

24. Install the tire/wheel assembly and lower the vehicle. ☐ Task completed

25. Once the car is set on the ground, tighten the new axle hub nut and wheel lug nuts to their specified torque. Install the wheel lug nuts and the new hub nut. What are the specifications?

____

26. Lower the vehicle. Tighten all bolts to specifications. Stake or use a cotter pin to keep the hub nut in place after it is tightened. What are the specifications?

____

27. Road test the car and recheck the torque on the hub nut. Summarize the results of the road test.

____

____

28. Check the vehicle's wheel alignment. Although indexing the alignment during disassembly will get the alignment angles close, they won't be at the desired settings. Camber can be off by as much as 3/4 degree due to differences between the size of the camber holes and the bolt. ☐ Task completed

Instructor's Comments

____

____

____

MANUAL TRANSMISSIONS JOB SHEET 22

Servicing Outer and Inner CV Joints

Name ______________________ Station ________________ Date ________________

NATEF Correlation

This Job Sheet addresses the following NATEF task:

C.4. Inspect, service, and replace shafts, yokes, boots, and CV joints.

Objective

Upon completion of this job sheet, you will be able to remove, inspect, and install both outer and inner CV joints.

Tools and Equipment

Brass drift
Miscellaneous hand tools
Rags
Service manual
Soft-jaw vise
Solvent tank or other cleaning equipment

Protective Clothing

Goggles or safety glasses with side shields

Describe the vehicle being worked on:

Year __________ Make ______________________ Model ______________________

VIN ______________________ Engine type and size ______________________

Describe general condition:

____

PROCEDURE (REMOVE AND INSPECT OUTER JOINT)

1. Remove the drive axle halfshaft from the vehicle, following the procedure in the service manual. ☐ Task completed
2. Secure the halfshaft in a soft-jaw vise. Mark the location of the boot to the shaft. Cut off the boot clamps and boot from the outer joint. Wipe away all grease to expose the snap ring or circlip. ☐ Task completed
3. If the joint is retained by a circlip, remove the joint from the shaft by striking the outer housing with a soft-faced hammer. ☐ Task completed
4. If the joint is retained by a snap ring, hold the snap ring open with snap-ring pliers and remove the joint from the shaft by striking the outer housing with a soft-faced hammer. ☐ Task completed

NOTE: *If the joint refuses to move, use a brass drift against the face of the inner joint race to drive the joint off the shaft.*

WARNING: *Always wear safety goggles or glasses with side shields when using a hammer and punch.*

5. Secure the outer housing in the vise to make working easier. Do not damage the splines in the outer housing end shaft with vise pressure. ☐ Task completed

6. Tilt the inner race and remove the joint balls in the sequence shown. ☐ Task completed

7. Tilt the inner race at a 90-degree angle to outer housing. Align the cage windows with the outer race, and then lift and remove the cage and inner race cross from the housing. Rotate the inner race upward and out of the cage. ☐ Task completed

8. Clean all components. ☐ Task completed

   NOTE: *After cleaning the constant velocity joint with solvent to inspect it, be sure to let the solvent dry completely before relubricating the reassembled joint. If the solvent is still wet, it might contaminate the lubricant and cause the reassembled CV joint to fail prematurely. It is also suggested that the CV joint be rinsed with a CV joint cleaner to remove residue completely.*

9. Inspect the joint for scoring or excessive wear on the bearings, cage windows, and inner and outer bearing races. Replace the entire joint if excessive wear is found on the bearings or grooves. Record the results on the Report Sheet for Inspection of Outer Constant Velocity Joints. ☐ Task completed

10. Examine the joint for cracks, chips, or brinelling. ☐ Task completed

    NOTE: *Brinelled cages are the main cause of clicking. A new cage with proper dimensions will normally correct this problem, but replacement of the entire wheel joint assembly is recommended if either condition exists. Shiny areas on the races and cages normally do not indicate a need for replacement unless the areas are excessively worn.*

    Record the results on the Report Sheet for Inspection of Outer Constant Velocity Joints.

PROCEDURE (PACK AND INSTALL JOINT)

1. If the components pass inspection, carefully reassemble the joint. Apply a light coat of oil to all parts before reassembly. ☐ Task completed

2. Align the inner race with the cage window and rotate it downward into the cage. Position the cage windows between the ball races and rotate them downward. ☐ Task completed

3 Swing the race cross and cage 90 degrees into the housing with the larger counterbore of the race cross facing outward. ☐ Task completed

4. Evenly distribute the grease provided in the repair kit into all ball bearing grooves. ☐ Task completed

5. Install as much grease as possible through the inner race and splined housing. Make sure that grease comes out between the outer race, cage, and bearings. Then install the rest of the grease into the boot. Use ALL of the grease supplied in the kit. ☐ Task completed

6. Tilt the cage and race cross (with the bearing grooves and windows aligned) toward the ball grooves in the housing, and insert the first ball. Insert the second ball the same way, but in the opposite side. Install the remainder of the balls in a similar fashion. ☐ Task completed

7. Loosely place the new boot clamp on the halfshaft. Carefully place the boot over the spline onto the shaft. Put electric tape over the splines to prevent boot damage during installation. ☐ Task completed

8. Install a new circlip. Position the joint on the shaft splines. Install the joint onto the shaft. Using a soft mallet, sharply tap the joint onto the splined halfshaft over the circlip. Pull lightly on the joint to make sure the circlip is seated correctly. ☐ Task completed

9. Install the boot and small clamp on the halfshaft. Use the remainder of the special grease to pack the boot. ☐ Task completed

10. With the large boot end properly placed on the joint housing, make sure no twists or crinkles appear on the boot. Install and tighten both boot clamps. ☐ Task completed

    NOTE: *The boot clamps on some vehicles must be tightened to a specified torque. Check the service manual before tightening.*

    CAUTION: *Do not overtighten.*

11. Flex the joints through a full range of motion to be sure they work smoothly. The complete halfshaft is now ready for installation. ☐ Task completed

PROCEDURE (REMOVE AND INSPECT INNER JOINT)

1. Turn the axle shaft so the inner joint can be worked on. ☐ Task completed

2. Push the housing to compress the retention spring lightly inside the joint. Bend the retaining tabs back using pliers. Hold the housing as the spring pushes it from the tripod. Separate the tripod and housing carefully. ☐ Task completed

3. Remove the snap ring. Do not allow the bearings to fall off the tripod. Use a wide elastic rubber band or tape to hold the bearings in place on the tripod. The tripod can be removed by hand or by tapping gently with a brass drift or wood block.. ☐ Task completed

    WARNING: *Always wear safety goggles or glasses with side shields when using a hammer and punch.*

4. Wipe any grease from the constant velocity (CV) joint bearings and bearing races. ☐ Task completed

    CAUTION: *Do not clean with solvent. Solvent could remain inside the needle bearing pockets and cause failure during operation.*

    Inspect the rest of the assembly for internal damage. Examine the fit between the rollers and housing. Excessive free play, roughness on either the roller or track surfaces, or damage to the bearings or trunnion call for joint replacement. Record the results on the Report Sheet for Inner Constant Velocity (CV) Joint.

5. Inspect the grease inside the housing. A gritty texture signals dirt penetration and possible damage. Clean the inside of the housing with rags. ☐ Task completed

PROCEDURE (JOINT INSTALLATION)

NOTE: *The replacement joint illustrated in this procedure is a spring-loaded design that does not have a retaining circlip to hold it in the differential.*

1. Wipe any remaining grease from the halfshaft splines and clean the shaft boot seating surface with solvent. Let the shaft dry completely. Temporarily tape the splines to avoid damaging the boot. ☐ Task completed
2. Slide the new boot onto the halfshaft and align the boot with the mark made before removing the old one. Install the boot clamp. Bend over the clamp and cut off all but 1/2 inch or so. Bend this small tap under itself with pliers, and lightly tap it secure. Remove the protective tape from the shaft. ☐ Task completed
3. Tape the rollers securely on the new tripod so they will not accidentally slide off the tripod during assembly. Place the tripod on the shaft splines with the chamfered or scribed end facing toward the outboard joint end. ☐ Task completed
4. Install the retaining ring into the shaft groove to lock the tripod assembly onto the shaft. Check that the ring is securely seated by pulling out the joint. ☐ Task completed
5. Pack the roller faces of the tulip housing raceway with the high-temperature grease contained in the replacement kit. Do not substitute any other type of grease. ☐ Task completed
6. Place the spring cup into the housing spring pocket, and lightly grease the cup. Reinstall the tulip housing. Slip the new retaining ring onto the shaft. Carefully remove the tape from the rollers and slide the tulip housing onto the tripod. Check that the spring remains in the pocket and that the cup connects with the round end of the connecting shaft. ☐ Task completed
7. Compress the spring by hand so the housing seats in the retaining ring. Use the properly sized C-clamp to hold the joint in a compressed position until the retaining ring can be secured. ☐ Task completed
8. To secure the retaining ring, remove the joint from the vise and use a hammer and punch to tightly compress the retaining ring to the housing. Work completely around the ring to ensure a tight fit. ☐ Task completed
9. Place the shaft back into the vise and position the boot on the tulip housing, making certain it is not dimpled or otherwise deformed. Remove any irregularities on the boot by working it between your fingers or by venting air into the boot with a dull screwdriver. ☐ Task completed
10. Install the large boot clamp and tighten it. Be careful not to cut or damage the boot or clamp as you work. ☐ Task completed

Problems Encountered

____

____

____

Instructor's Comments

____

____

____

REPORT SHEET FOR INSPECTION OF OUTER CONSTANT VELOCITY JOINTS		
	Serviceable	*Nonserviceable*
Visual Inspection		
Balls		
Cage		
Inner race		
Outer race		

Conclusions and Recommendations ____

____

____

REPORT SHEET FOR INNER CV JOINT		
	Serviceable	*Nonserviceable*
Inspection		
Rollers		
Needle bearings		
Trunnion		
Outer "tulip" cage		

Conclusions and Recommendations ______________________________

____

____

MANUAL TRANSMISSIONS JOB SHEET 23

Servicing the Center Support Bearing

Name ______________________ Station ______________ Date ________________

NATEF Correlation

This Job Sheet addresses the following NATEF task:

C.5. Inspect, service, and replace shaft center support bearings.

Objective

Upon completion of this job sheet, you will be able to inspect, service, and replace driveshaft center support bearings.

Tools and Materials

Clean rags

Hand tools

Protective Clothing

Goggles or safety glasses with side shields

Describe the vehicle being worked on:

Year _________ Make _____________________ Model ____________________

VIN _____________________ Engine type and size _________________________

PROCEDURE

1. Does the vehicle have a two-piece driveshaft to the rear axle? __________
2. If so, carefully inspect the center bearing assembly for damage or signs of excessive wear. Summarize your findings.

   ____

   ____
3. Grab the two driveshafts that are connected by the center bearing and move them in all directions, looking for signs of wear or looseness. Summarize your findings.

   ____

   ____
4. If the center bearing is worn or damaged, it must be replaced. To do so, remove the bolts or nut that retains the rear driveshaft to the flange at the rear axle. ☐ Task completed
5. Pull the driveshaft out of the center bearing. ☐ Task completed

6. Loosen and remove the mounting bolts for the center bearing. ☐ Task completed
7. Support the front half of the driveshaft. ☐ Task completed
8. Remove the center bearing assembly. ☐ Task completed
9. Install the new center bearing assembly. Make sure you properly tighten the bolts or nuts. ☐ Task completed
10. Insert the rear portion of the driveshaft into the bearing. ☐ Task completed
11. Reconnect the driveshaft to the rear axle flange. ☐ Task completed

Instructor's Comments

___

___

___

MANUAL TRANSMISSIONS JOB SHEET 24

Drive Axle Leak Diagnosis

Name ______________________________ Station ________________ Date ________________

NATEF Correlation

This Job Sheet addresses the following NATEF task:

D.1.2. Diagnose fluid leakage concerns; determine necessary action.

Objective

Upon completion of this job sheet, you will be able to identify the cause of drive axle fluid leakage problems.

Tools and Materials

Basic hand tools

Protective Clothing

Goggles or safety glasses with side shields

Describe the vehicle being worked on:

Year __________ Make ________________________ Model ________________________

VIN ________________________ Engine type and size ________________________

PROCEDURE

1. To identify the exact source for a leak, a careful inspection must be completed. Look at the drive axle and note the spots where there may be a leak.

   ____

   ____

2. Check the area around the pinion shaft. An improperly installed or damaged drive pinion seal will allow the lubricant to leak past the outer edge of the seal. Any damage to the seal's bore, such as dings, dents, and gouges, will distort the seal casing and allow leakage. Also, the spring that holds the seal lip against the companion flange may be knocked out and allow leakage past the seal's lip. Clean the area around the leak so you can determine the location of the leak and record your findings.

   ____

   ____

3. It is also possible for oil to leak past the threads of the drive pinion nut or the pinion retaining bolts. Removing the nut or bolts, applying thread or gasket sealer on the threads, and torquing the nuts or bolts to specifications can stop these leaks. Clean the area around the leak so you can determine the location of the leak and record your findings.

   ____

   ____

4. Leakage past the stud nuts that hold the carrier assembly to the axle housing can be corrected by installing copper washers under the nuts. Always make sure there is a copper washer under the axle ID tag. Clean the area around the leak so you can determine the location of the leak and record your findings.

____

____

5. Check the housing cover for signs of leakage. Check for poor gasket installation, loose retaining bolts, or a damaged mating surface. Most late-model vehicles do not use a gasket on the housing cover; instead, silicone sealer is used. The old sealant should be cleaned off before applying a new coat to the surface. If a housing cover is leaking, inspect the surface for imperfections, such as cracks or nicks. File the surface true and install a new gasket. If the surface cannot be trued, apply some gasket sealer to the surface before installing the new gasket. Always follow the recommended tightening sequence and torque specifications when tightening an axle housing cover. Describe your findings.

____

____

6. Check for lubricant leaks on the housing itself. At times, lubricant will leak through the pores of the housing. To correct this problem when the porous area is small, force some metallic body filler into the area. After the filler has set, seal the area with an epoxy-type sealer. If the leaking pore is rather large, drill a hole through the area and tap an appropriately sized setscrew into the hole, then cover the area with an epoxy sealer. Describe your findings.

____

____

7. Check the level of the fluid in the assembly. Describe your findings and state what type of fluid is recommended for this vehicle.

____

____

____

8. Check the vent for signs of leakage. If the wrong vent was installed in the axle housing or if there is an excessive amount of lubricant or oil turbulence in the axle, lubricant may leak through the axle vent hose. Also check for a crimped or broken vent hose. If the cause of the leakage is an overfilled axle assembly, drain the housing and refill the unit with the specified amount and type of lubricant. Describe your findings.

____

____

9. Check the brake backing plates and drums for signs of fluid. If fluid is present, it is usually caused by a worn or damaged axle seal. However, if the seal's bore in the axle tube is damaged, the seal will be unable to seal properly. Whenever fluid is leaking past the axle seal, the bore should be checked and the seal replaced. Describe your findings.

____

____

10. Some drive axle units are fitted with an ABS sensor. Lubricant can leak from around the sensor's O-ring if it is damaged. Check here and describe your findings.

____

____

Instructor's Comments

____

____

____

MANUAL TRANSMISSIONS JOB SHEET 25

Companion Flange and Pinion Seal Service

Name ______________________ Station ______________ Date ______________

NATEF Correlation

This Job Sheet addresses the following NATEF task:

D.1.3. Inspect and replace companion flange and pinion seal; measure companion flange runout.

Objective

Upon completion of this job sheet, you will be able to inspect and replace the companion flange and pinion seal. You will also be able to measure companion flange runout.

Tools and Materials

Pinion flange holding tool
Catch pan
Center punch
Slide hammer
Fresh lubricant
Seal remover
Seal puller
Inch-pound torque wrench
Basic hand tools

Protective Clothing

Goggles or safety glasses with side shields

Describe the vehicle being worked on:

Year ________ Make __________________ Model __________________

VIN __________________ Engine type and size __________________

PROCEDURE

1. If the pinion seal is a source of fluid leakage, it should be replaced. To replace a pinion seal, remove the drive shaft from the companion flange. ☐ Task completed
2. Remove the flange nut. You may need a special wrench to hold the flange while removing the nut. ☐ Task completed
3. Pull the companion flange off the pinion splines. ☐ Task completed
4. Inspect the sealing surface of the flange to make sure it is smooth and free of burrs. What did you find?

   ____

   ____
5. With a slide-hammer, remove the pinion seal. ☐ Task completed

6. Before you install a new seal, lubricate the seal's lip and coat the outside diameter of the seal with gasket sealer. Use a seal driver to install the new seal flush against the carrier or axle housing. Make sure the spring around the seal lip remains in place. ☐ Task completed

7. After the seal is installed, lubricate the seal area of the companion flange, and push it carefully into the seal, making sure the seal isn't damaged. ☐ Task completed

8. Start a new pinion flange nut onto the threads and set bearing preload. The tightness of the pinion flange nut is critical because it determines the pinion bearing preload. If the flange nut is overtightened, the pinion bearing preload must be readjusted. Describe the recommended procedure for setting bearing preload.

____

____

____

____

9. Excessive companion flange runout can cause a number of problems, including wear on the seal and driveline vibrations. To check flange runout, mount a dial indicator on the axle housing so that the indicator's plunger rests on the outside face of the flange. ☐ Task completed

10. Mark the spot where the indicator's plunger contacts the flange. ☐ Task completed

11. Set the indicator to zero and rotate the wheels to turn the flange. Rotate the flange one complete turn. ☐ Task completed

12. The total amount of needle deflection on the indicator is the total amount of runout. What was the measured runout?

____

____

13. Compare your reading to the specifications and summarize the results.

____

____

____

____

14. Once the flange nut is installed and properly tightened, install the drive shaft. ☐ Task completed

15. Check the fluid in the differential, then road test the vehicle and again check for leaks. ☐ Task completed

Instructor's Comments

____

____

____

MANUAL TRANSMISSIONS JOB SHEET 26

Measure and Adjust Pinion Depth, Bearing Preload, and Backlash

Name ______________________________ Station __________________ Date __________________

NATEF Correlation

This Job Sheet addresses the following NATEF tasks:

D.1.4. Inspect ring gear and measure runout; determine necessary action.

D.1.6. Measure and adjust drive pinion depth.

D.1.7. Measure and adjust drive pinion bearing preload.

D.1.8. Measure and adjust side bearing preload and ring and pinion gear total backlash and backlash variation on a differential carrier assembly (threaded cup or shim types).

D.1.9. Check ring and pinion tooth contact patterns; perform necessary action.

D.1.10. Disassemble, inspect, measure, and adjust or replace differential pinion gears (spiders), shaft, side gears, side bearings, thrust washers, and case.

Objective

Upon completion of this job sheet, you will be able to check and adjust drive pinion depth, bearing preload, and ring and pinion gear backlash.

Tools and Equipment

Hand tools
Dial indicator
Pinion depth gauge
Torque wrench
Service manual
Pinion gear flange holding tool
Micrometer

Protective Clothing

Goggles or safety glasses with side shields

Describe the vehicle being worked on:

Year __________ Make ________________________ Model ________________________

VIN ________________________ Engine type and size ________________________

PROCEDURE (DRIVE PINION DEPTH)

1. Check the condition of the pinion bearings; replace them if necessary. ☐ Task completed
2. Check the pinion gear for any markings indicating additional adjustments.
   Record the markings: ____
3. Set up the pinion depth gauge according to the procedure outlined in the service manual. ☐ Task completed

4. Set up the dial indicator on the carrier housing. ☐ Task completed
5. Make the necessary readings with the indicator and record the results: ☐ Task completed
6. How much needs to be added or subtracted to achieve proper pinion depth? ____________
7. Refer to the service manual to determine the correct size of pinion shim that should be used. ☐ Task completed
8. Install the shim and bearing on the pinion gear shaft. ☐ Task completed
9. Install the pinion gear into the carrier housing. ☐ Task completed

PROCEDURE (PINION BEARING PRELOAD)

1. Install the pinion gear, crush sleeve, and bearing into the carrier housing. ☐ Task completed
2. Install the pinion seal into the housing. ☐ Task completed
3. Install the pinion flange, washer, and nut on the pinion. ☐ Task completed
4. Using the flange holding tool, tighten the pinion nut. ☐ Task completed
5. Using a torque wrench, measure the torque required to turn the pinion gear. Required torque is: ____________ ☐ Task completed
6. Refer to the service manual for the proper torque required to turn the pinion. Specified torque is: ____________ ☐ Task completed
7. Tighten the pinion nut until the proper torque reading is reached. ☐ Task completed

PROCEDURE (RING AND PINION GEAR BACKLASH)

1. Check the ring gear for runout by setting the dial indicator on the back side of the ring gear.
2. Rotate the ring gear one complete revolution and note the movement on the dial indicator. Describe what you observed:

   ____

   ____
3. What was the highest reading? ____________
4. What was the lowest reading? ____________
5. Subtract the lowest from the highest; this indicates the total runout of the ring gear. What was it? ____________
6. If the runout was not within specifications, check the runout of the carrier before replacing the ring gear. ☐ Task completed
7. Now, install the differential case and ring gear into the carrier housing. ☐ Task completed

8. Mount the dial indicator onto the carrier housing. ☐ Task completed
9. Set the dial indicator on a ring gear tooth. ☐ Task completed
10. Look up the specifications for backlash and record them here:

    ____

11. Rock the ring gear back and forth against the teeth of the pinion gear. ☐ Task completed
12. Observe the total movement of the indicator; this is the total backlash. Your readings were: ______________________
13. Measure backlash at four different spots on the ring gear. ☐ Task completed
14. Describe what needs to happen to correct the backlash:

    ____

    ____

    ____

15. Using knock-in shims or adjusting nuts (depending on axle design), move the ring gear in reference to the pinion gear to achieve proper backlash. ☐ Task completed
16. When proper backlash is reached, torque the retaining caps to specifications. ☐ Task completed
17. Recheck the backlash to make sure it is still within specifications. ☐ Task completed

Problems Encountered

____

____

____

Instructor's Comments

____

____

____

MANUAL TRANSMISSIONS JOB SHEET 27

Servicing the Ring and Pinion Gears

Name ______________________ Station _______________ Date _________________

NATEF Correlation

This Job Sheet addresses the following NATEF task:

D.1.5. Remove, inspect, and reinstall drive pinion and ring gear, spacers, sleeves, and bearings.

Objective

Upon completion of this job sheet, you will be able to remove, inspect, and reinstall the drive pinion and ring gears, spacers, sleeves, and bearings.

Tools and Materials

Paint stick or marking tool
Pinion flange holding tool
Soft-faced hammer
Various drivers
Hydraulic press
Micrometer

Protective Clothing

Goggles or safety glasses with side shields

Describe the vehicle being worked on:

Year _________ Make _____________________ Model ____________________

VIN _____________________ Engine type and size _________________________

PROCEDURE

1. With the carrier removed, mark the alignment of the ring gear to the differential case. ☐ Task completed
2. Check the ring gear bolts for markings that may indicate they are left-handed threads. Did you observe any?

   ____

3. Loosen and remove the ring gear to differential case bolts. ☐ Task completed
4. Loosen and remove the pinion retaining nut. ☐ Task completed
5. Drive the pinion shaft out of the front bearing with a soft-faced hammer and remove it from the housing. ☐ Task completed
6. Drive the front bearing cup out of the housing and remove the front seal. ☐ Task completed
7. Using the correct fixture, press the pinion bearing off the shaft. Be sure to remove, measure, and record the thickness of the shim located behind the bearing. The thickness of the shim is _____________.

8. Secure the new ring gear and position it on the differential case. Did you need to heat up the gear to install it? ____________

9. Install new ring gear bolts and tighten them to specs. What were the torque specifications for these bolts? ________________________

10. Determine the correct size shim to install on the pinion shaft to correctly place the pinion gear to the ring gear. How did you determine the size?

____

____

____

11. Using the correct fixture, press the pinion bearing onto the pinion shaft. Make sure you installed the correct shim behind the bearing. ☐ Task completed

12. Install the pinion shaft into the housing. ☐ Task completed

13. Install the pinion seal, flange, and nut. Tighten the nut to specifications. What is the torque spec for this nut? ______________

14. Measure and set pinion depth with the appropriate tools. What did you use?

____

____

15. The differential case is now ready for installation into the housing. Make sure you check gear backlash and side bearing side preload. ☐ Task completed

Instructor's Comments

____

____

____

MANUAL TRANSMISSIONS JOB SHEET 28

Differential Case Service

Name ______________________ Station ______________ Date ________________

NATEF Correlation

This Job Sheet addresses the following NATEF task:

D.1.11. Reassemble and reinstall differential case assembly; measure runout; determine necessary action.

Objective

Upon completion of this job sheet, you will be able to reassemble and reinstall a differential case assembly and measure runout.

Tools and Materials

Dial indicator
Differential holding tool
Brass drift
Gear/bearing puller
Hydraulic press
Hand tools

Protective Clothing

Goggles or safety glasses with side shields

Describe the vehicle being worked on:

Year _________ Make _____________________ Model ____________________

VIN _____________________ Engine type and size ________________________

PROCEDURE

1. With the differential unit still in the housing, set a dial indicator so that its plunger rides on the differential case. ☐ Task completed
2. By hand, rotate the case and observe the readings on the dial indicator. What did you see during one complete revolution of the case?

____

____

3. Based on the above, what are your recommendations?

____

____

4. If the differential case runout was okay, put the case in a holding fixture. ☐ Task completed
5. Remove and discard the ring gear bolts. ☐ Task completed

6. With a brass drift, tap the ring gear loose from the differential case. ☐ Task completed
7. Rotate the side gears until the pinion gears appear at the case window. ☐ Task completed
8. Remove the pinion gears, side gears, and thrust washers. ☐ Task completed
9. Using the appropriate puller, remove the side bearings. ☐ Task completed
10. To begin reassembly, install the thrust washers on the differential side gears. ☐ Task completed
11. Position the side gears in the differential case. ☐ Task completed
12. Place the thrust washers on the pinion gears, then mesh the pinion gears with the side gears. ☐ Task completed
13. Install the differential pinion shaft, and align and install the lock pin into the pinion shaft. ☐ Task completed
14. Heat the ring gear and install the ring gear onto the differential case. ☐ Task completed
15. Install and tighten the new ring gear bolts. ☐ Task completed
16. Use the correct driver and press to install the differential case side bearings. ☐ Task completed
17. Install the differential with its bearing cones and proper shims. ☐ Task completed
18. Install the bearing caps and tighten to specifications. What are the specifications? ______________________
19. Check the preload and backlash of the gear set. ☐ Task completed
20. Install the axles, C-locks, differential pinion shaft lock pin, and the rear cover. ☐ Task completed
21. Install the brake drums (or rotors), wheels, and driveshaft. ☐ Task completed
22. Refill the axle housing with the appropriate type and amount of fluid. What kind of fluid did you put in? How much of it did you put in?

____

____

____

Instructor's Comments

____

____

____

MANUAL TRANSMISSIONS JOB SHEET 29

Limited-Slip Differential Diagnostics

Name ______________________ Station ______________ Date ______________

NATEF Correlation

This Job Sheet addresses the following NATEF tasks:

D.2.1. Diagnose noise, slippage, and chatter concerns; determine necessary action.

D.2.4. Measure rotating torque; determine necessary action.

Objective

Upon completion of this job sheet, you will be able to diagnose noise, slippage, and chatter concerns in a limited-slip differential assembly. You will also be able to measure the rotating torque of the assembly.

Tools and Materials

Axle shaft puller

Torque wrench

Basic hand tools

Protective Clothing

Goggles or safety glasses with side shields

Describe the vehicle being worked on:

Year ________ Make ____________________ Model ____________________

VIN ____________________ Engine type and size ____________________

PROCEDURE

1. Take the vehicle for a road test. Pay attention to the noise level and action of the differential unit as you make a right and left turn. Describe your findings.

   ____

   ____

2. Now drive the vehicle on a slick surface and accelerate hard enough to lose traction. Pay attention to the noise level and action of the differential unit. Describe your findings.

   ____

   ____

3. Based on the above tests, is the unusual noise or action caused by the limited slip function of the differential or is it caused by something else? Explain your answer.

   ____

   ____

4. A limited-slip differential can be checked for proper operation without removing the differential from the axle housing. Be sure the transmission is in neutral, one rear wheel is on the floor, and the other rear wheel is raised off the floor. Specifications should give the breakaway torque reading required to start the rotation of the wheel that is raised off the floor. What are the specifications?

____

____

5. Using a torque wrench, measure the torque required to turn the wheel. What was your measurement and how does it compare with the specification?

____

____

Instructor's Comments

____

____

____

MANUAL TRANSMISSIONS JOB SHEET 30

Differential Housing Service

Name ______________________ Station ______________ Date ______________

NATEF Correlation

This Job Sheet addresses the following NATEF task:

D.2.2. Inspect and flush differential housing; refill with correct lubricant.

Objective

Upon completion of this job sheet, you will be able to inspect and flush a differential housing and refill it with the correct lubricant.

Tools and Materials

Basic hand tools

Clean rag

Protective Clothing

Goggles or safety glasses with side shields

Describe the vehicle being worked on:

Year ________ Make ____________________ Model ____________________

VIN ____________________ Engine type and size ____________________

Transmission type and number of forward speeds ________________________

PROCEDURE

1. On a RWD differential and final drive assembly, check the fluid level with the vehicle on a level surface and the axle at normal operating temperature. The fluid level should be even with the bottom of the fill plug opening in the axle housing unless otherwise specified. State the type of recommended fluid.

   ____

   ____

2. If the fluid level is low, inspect the differential and axle housing for signs of leakage. What did you find?

   ____

   ____

3. State the condition of the fluid and explain what you think may have caused the unusual smell or the contamination.

   ____

   ____

4. On most FWD models with manual transaxles, the transmission and final drive assembly are lubricated with the same fluid and share the same fluid reservoir. State the type of recommended fluid.

____

____

5. Check the fluid level at the transaxle fill plug or dipstick. The fluid should be even with the bottom of the fill plug opening unless otherwise specified. What did you find?

____

____

6. State the condition of the fluid and explain what you think may have caused the unusual smell or the contamination.

____

____

7. If the fluid was contaminated and the problem was not caused by the destruction of parts in the differential, the housing should be flushed. To do this, drain the fluid from the housing. ☐ Task completed

8. Then refill the housing with clean fluid and rotate the wheels and/or drive shaft to circulate the fluid. ☐ Task completed

9. Drain this fluid and refill the housing with a fresh supply of fluid. ☐ Task completed

10. Rotate the wheels and/or drive shaft several times, and then reexamine the condition of the fluid. If the fluid still has traces of residue, flush the housing again or until the fluid is clean. ☐ Task completed

11. Is the differential a limited-slip unit? If so, what fluid is recommended and what additive should you mix with the fluid? Before adding friction modifiers to the lubricant of a limited-slip differential, always check its compatibility. Using the wrong additive may cause the axle's lubricant to break down, which would destroy the axle very quickly.

____

____

Instructor's Comments

____

____

____

MANUAL TRANSMISSIONS JOB SHEET 31

Servicing Limited-Slip Differentials

Name ______________________ Station ______________ Date ______________

NATEF Correlation

This Job Sheet addresses the following NATEF task:

D.2.3. Inspect and reinstall clutch (cone or plate) components.

Objective

Upon completion of this job sheet, you will be able to inspect and reinstall clutch components in a limited-slip differential.

Clutch gauge kit Torque wrench
Feeler gauge Basic hand tools

Protective Clothing

Goggles or safety glasses with side shields

Describe the vehicle being worked on:

Year ________ Make ____________________ Model ____________________

VIN ____________________ Engine type and size ____________________

PROCEDURE

NOTE: *Some limited-slip units (such as the Auburn, Torsen, and many with viscous clutches) are not serviceable. When these units fail, they are replaced as a unit.*

1. Disassemble the clutch pack. Inspect each friction and steel plate as you are removing them. Describe the condition of each.

   ____

   ____

2. Check the condition of the preload spring and all springs used in the unit. Any distortion is justification for replacing them during reassembly. Describe the condition of each.

   ____

   ____

3. Inspect the side gear retainers for excessive wear and cracks. ☐ Task completed

4. Identify all parts that need to be replaced and secure those replacement parts. List the parts that are to be replaced. ☐ Task completed

5. Assembly is the reverse of disassembly. Make sure the ears of the steel plates fit into the case. ☐ Task completed

6. Place the composite plate so that it is positioned with its fiber side against the hub. The other plates are placed in alternating order, with a friction disc followed by a steel one. ☐ Task completed

7. Clutch packs should be allowed to soak in gear lube for at least 30 minutes prior to installation and before doing the following check. ☐ Task completed

8. Check the total width of the clutch pack with a template or other suitable gauge and a feeler gauge to determine the correct thickness of the shim needed to maintain preload. Always refer to your shop manual to determine the proper way to measure the clutch pack preload. How did you check the preload and how did the measured preload compare to specifications?

____

____

9. An additional check should be made on the unit to determine if it is working properly. Use a torque wrench and measure the torque required to rotate one side gear while the other is held stationary. Compare this reading of initial breakaway torque to the specifications given in your service manual. What are the specifications and the results of this check?

____

____

Instructor's Comments

____

____

____

MANUAL TRANSMISSIONS JOB SHEET 32

Axle Shaft and Bearing Service

Name ______________________ Station ______________ Date ______________

NATEF Correlation

This Job Sheet addresses the following NATEF tasks:

D.3.1. Diagnose drive axle shafts, bearings, and seals for noise, vibration, and fluid leakage concerns; determine necessary action.

D.3.2. Inspect and replace drive axle shaft wheel studs.

D.3.3. Remove and replace drive axle shafts.

D.3.4. Inspect and replace drive axle shaft seals, bearings, and retainers.

D.3.5. Measure drive axle flange runout and shaft endplay; determine necessary action.

Objective

Upon completion of this job sheet, you will be able to diagnose, inspect, remove, and replace axle shafts, bearings, seals, and retainers. You will also be able to inspect and replace drive axle wheel studs and measure drive axle flange runout and shaft endplay.

Tools and Materials

Vee blocks | Axle puller
Dial indicator | Slide hammer
Cold chisel | Bearing adapters

Protective Clothing

Goggles or safety glasses with side shields

Describe the vehicle being worked on:

Year ________ Make __________________ Model __________________

VIN __________________ Engine type and size __________________

PROCEDURE

1. The diagnosis of any problem should begin with a detailed conversation with the customer to gather as much information about the problem as possible. Ask the customer to carefully describe the problem. This description should include any noises or vibrations, when the problem is evident, and when the problem first became noticeable. If the customer's complaint is based on an abnormal noise or vibration, find out where it is felt or heard. You should also find out if the problem resulted from an event or mishap, such as running into a curb or pothole. Ask the customer about the service history of the car; often drive axle problems can be caused by mistakes made by technicians while servicing the vehicle. What are the customer's concerns?

____

____

____

2. If it is possible, take the customer along with you on the road test. Attempt to duplicate the customer's complaint by operating the vehicle under conditions like those during which the problem occurs. Pay careful attention to the vehicle during those and all operating conditions. Keep notes of the vehicle's behavior while driving under various conditions. Note the engine and vehicle speeds at which the problem is most evident. What were the results of the road test?

____

____

____

3. While diagnosing a drive axle noise, it is important to operate the vehicle in all possible conditions and note any changes in noise during each mode. It is also helpful to note how a change in speed affects the noise or vibration. Before assuming that the drive axle is the source of a noise or vibration, make certain the tires or exhaust are not the cause. Inspect the exhaust, suspension, and tires for possible causes of the noise and summarize your findings.

____

____

4. Poor lubrication in the axle can also be a source of noise, so the lubricant's level and condition should be checked. ☐ Task completed

5. If the level is low, the axle should be filled with the proper type and amount of lubricant and the vehicle road tested again. If the lubricant is contaminated, it should be drained and the axle refilled. Low or contaminated lubricant in a drive axle indicates a leak in the assembly. Therefore, the assembly should be carefully inspected to locate the source of the leak. What did you find?

____

____

____

Checking the Axle Flange

1. Axle shaft replacement is required if there is excessive runout of the axle's flange. To check the runout of the flange, position a dial indicator against the outer flange surface of the axle. ☐ Task completed

2. Apply slight pressure to the center of the axle to remove the endplay in the axle, then zero the indicator. ☐ Task completed

3. Slowly rotate the axle one complete revolution and observe the readings on the indicator. The total amount of indicator movement is the total amount of axle flange lateral runout. Compare the measured runout with specifications and summarize your findings.

____

____

4. Inspect the wheel studs in the axle flanges. If they are broken or bent, they should be replaced. What did you find?

____

____

5. Also check the condition of the threads. If they have minor distortions, run a die over the stud. If the threads are severely damaged, the stud should be replaced. Studs are normally pressed in and out of the flange. Make sure you install the correct size stud. What did you find?

____

____

Axle Shaft Service

1. If the axle shaft is held in by a retainer, it is not necessary to remove the differential cover to remove the axle. To remove the axle, remove the four bolts that hold the retainer to the backing plate, then pull the axle out. Normally the axle shaft will slide out without the aid of a puller, but sometimes a puller is required. Did you face any difficulties?

____

____

2. A tapered-roller bearing supported axle shaft is secured by a C-lock located inside the differential, and the differential cover must be removed to gain access to it. To remove this type of axle, first remove the wheel, brake drum, and differential cover. Then remove the differential pinion shaft retaining bolt and differential pinion shaft. Now push the axle shaft in and remove the C-lock. The axle can now be pulled out of the housing. Did you face any difficulties?

____

____

3. When an axle is removed from a housing, the bearing's bore in the axle housing should be inspected and cleaned. If the bore is corroded, sand it lightly with fine emery cloth, then coat the bore with non–water-soluble grease. This grease will allow the bearing to move slightly in the housing, thereby increasing its service life. The grease will also help prevent corrosion of the bearing's outer race. What did you find?

____

____

4. Place the shaft in a pair of Vee-blocks and position a dial indicator at the center of the shaft. Rotating the shaft and observing the indicator will identify any warping or bends in the shaft. Slight indications on the dial indicator may be caused by casting imperfections and do not indicate the need to replace the shaft. What did you find?

____

____

5. Ball-type bearings are lubricated with grease packed in the bearing at the factory. This type of bearing is pressed on and off the axle shaft. The retainer ring is made of soft metal and is pressed onto the shaft against the wheel bearing. The ring can be slid off the shaft after it has been drilled into or notched with a cold chisel to break the seal. Did you face any difficulties?

____

____

6. Roller axle bearings are lubricated by the gear oil in the axle housing. These bearings are typically pressed into the axle housing. To remove them, the axle must first be removed and then the bearing pulled out of the housing. With the axle out, inspect the area where it rides on the bearing for pits or scores. If pits or score marks are present, replace the axle. What did you find?

___

___

7. Tapered-roller axle bearings must be pressed on and off the axle shaft using a press. After the bearing is pressed onto the shaft, it must be packed with wheel bearing grease. After packing the bearing, install the axle in the housing and check the shaft's endplay. If the endplay is not within specifications, change the size of the bearing shim. Did you face any difficulties?

___

___

8. All axle bearings should be inspected for wear or other damage. If the bearings show any evidence of damage, they should be replaced. What did you find?

___

___

9. The installation of new axle shaft seals is recommended whenever the axle shafts have been removed. Some axle seals are identified as being either right or left side. When installing new seals, make sure to install the correct seal in each side. How are your seals marked?

___

___

10. Coat the outer edge of the new seal with gasket sealer and apply a coat of lubricant to the seal's inner lip. Make sure the lip seal is facing the inside of the axle housing, then use the correct size seal driver to install the seal squarely in the axle tube. Did you face any difficulties?

___

___

11. In most cases, the installation of axle shafts is a simple procedure. Prior to installing the shafts, make sure you installed all of the bearings, seals, and retaining plates on the shaft. ☐ Task completed

Axle Endplay

1. Some axles with tapered-roller bearings require an endplay adjustment. To check the endplay, position the dial indicator so it can measure the end-to-end movement of the axle. ☐ Task completed

2. Push the axle into the housing and set the indicator to zero. ☐ Task completed

3. Pull on the axle and note the indicator's reading. This is the amount of endplay. Compare the reading against specifications and correct the endplay as necessary. What were the results?

___

___

4. Endplay adjustments are made by adding or subtracting shims or by turning the adjusting nut. This adjustment is done on one side of the housing but sets the endplay for both sides. What did you need to do to correct the endplay?

____

____

____

Instructor's Comments

____

____

____

MANUAL TRANSMISSIONS JOB SHEET 33

Road Test a Transfer Case

Name ______________________ Station ______________ Date ______________

NATEF Correlation

This Job Sheet addresses the following NATEF task:

E.1. Diagnose noise, vibration, and unusual steering concerns; determine necessary action.

Objective

Upon completion of this job sheet, you will be able to road test a vehicle and determine the operating condition of a transfer case.

Tools and Equipment

Vehicle with four-wheel-drive

Describe the vehicle being worked on:

Year ________ Make ____________________ Model ____________________

VIN ____________________ Engine type and size ____________________

Describe the transfer case controls and list the type of transfer case found on the vehicle:

____

PROCEDURE

Road test the vehicle and attempt to operate the vehicle in all of the modes of the transfer case.

1. Does the transfer case make noise in:
   a. Low drive rear-wheel drive? _____ yes _____ no _____ n/a
   b. High drive in rear-wheel drive? _____ yes _____ no _____ n/a
   c. Low drive in four-wheel drive? _____ yes _____ no _____ n/a
   d. High drive in four-wheel drive? _____ yes _____ no _____ n/a
2. Does it make noise as you turn corners with it in four-wheel drive? _____ yes _____ no
3. Does it make noise as you drive down the road in high gear? _____ yes _____ no
4. Does it make a noise when the transfer case is in neutral? _____ yes _____ no

Problems Encountered

Instructor's Comments

MANUAL TRANSMISSIONS JOB SHEET 34

Servicing Transfer Case Shift Controls

Name ______________________ Station ______________ Date _________________

NATEF Correlation

This Job Sheet addresses the following NATEF task:

E.2. Inspect, adjust, and repair shifting controls (mechanical, electrical, and vacuum), bushings, mounts, levers, and brackets.

Objective

Upon completion of this job sheet, you will be able to inspect, adjust, and repair the shifting controls (including mechanical, electrical, and vacuum), as well as the bushings, mounts, levers, and brackets for a transfer case.

Tools and Materials

DMM
Wiring diagram
Vacuum gauge
Hand-operated vacuum pump
Service manual

Protective Clothing

Goggles or safety glasses with side shields

Describe the vehicle being worked on:

Year _________ Make _____________________ Model ____________________

VIN _____________________ Engine type and size ________________________

PROCEDURE

1. Raise the vehicle on a lift. ☐ Task completed

2. Locate and check the mechanical linkages at the transfer case for looseness and damage. Record your findings here.

   ____

   ____

3. Carefully check the linkage bushings and brackets. Record your findings here.

   ____

   ____

4. Check the mounts for the transfer case and record your findings here.

____

____

5. If the mechanical linkage is fine or if there is no direct linkage, proceed to check the electrical controls by lowering the vehicle.

☐ Task completed

6. Locate the shift controls and circuit on the wiring diagram. What main components are part of the circuit?

____

____

7. Check the control switch for high resistance, which could be the cause of poor shifting. How did you test the switch and what did you find?

____

____

8. With the ignition on, use the switch to move from 2WD to 4WD. You should have heard the click of the solenoid or motor. If the clicking was heard, what is indicated?

____

____

9. If there was no click, check the circuit fuse. If the fuse is bad, there must be a short in the circuit. What was the condition of the fuse?

____

____

10. Raise the vehicle and check for power at the solenoids. Record your findings here.

____

____

11. What would be indicated by having the solenoids not working after voltage was applied to them?

____

____

12. Visually inspect all electrical connections and terminals. Record your results.

____

____

13. If the electrical circuit is fine, check engine manifold vacuum. Record your results and summarize what the results indicate.

____

____

14. Visually inspect the vacuum lines for tears, looseness, and other damage. Record your findings here.

____

____

15. With the vacuum pump, create a vacuum at the servo and observe its activity and ability to hold a vacuum. Record your results here.

____

____

16. If the system has a vacuum motor, apply a vacuum to it and observe its activity and ability to hold a vacuum. Record your results here.

____

____

17. Based on all these checks, what are your service recommendations for this vehicle?

____

____

____

Instructor's Comments

____

____

____

MANUAL TRANSMISSIONS JOB SHEET 35

Removing and Installing a Transfer Case

Name ______________________ Station _______________ Date __________________

NATEF Correlation

This Job Sheet addresses the following NATEF task:

E.3. Remove and reinstall a transfer case.

Objective

Upon completion of this job sheet, you will be able to remove and reinstall a transfer case.

Tools and Materials

Drain pan
Clean rags
Transmission jack
Hand tools
Service manual

Protective Clothing

Goggles or safety glasses with side shields

Describe the vehicle being worked on:

Year _________ Make ______________________ Model ____________________

VIN ______________________ Engine type and size _________________________

PROCEDURE

NOTE: *These are intended to be basic guidelines. Refer to the service manual and modify the following sequence according to the procedures outlined by the manufacturer.*

1. Raise the vehicle on a lift. ☐ Task completed
2. Place a drain pan under the transfer case and remove the drain plug. ☐ Task completed
3. Disconnect and label all wires connected to the transfer case. ☐ Task completed
4. Disconnect and remove the front driveshaft from the transfer case. ☐ Task completed
5. Disconnect and remove the rear driveshaft from the transfer case. ☐ Task completed
6. Disconnect the speedometer cable if it is attached to the transfer case. ☐ Task completed
7. Disconnect the vent hose from the case. ☐ Task completed
8. Reinsert the drain plug and move the drain pan away from the vehicle. ☐ Task completed
9. Support the transfer case with the transmission jack. ☐ Task completed

10. Remove the bolts that attach the transfer case to the transmission or to a support. ☐ Task completed

11. Slide the transfer case rearward and remove it from the vehicle. ☐ Task completed

12. Before reinstalling the transfer case, make sure the mounting surface on the transmission is clean and all old gasket material is removed. ☐ Task completed

13. Look up all installation torque specs and record them here.

____

____

14. With the transmission jack, raise the transfer case into position. ☐ Task completed

15. Install a new gasket at the mounting surface and slide the transfer case forward until it is fully seated on the transmission or support. ☐ Task completed

16. Install and tighten the bolts that attach the transfer case to the transmission or to the support. ☐ Task completed

17. Move the transmission jack away from the vehicle. ☐ Task completed

18. Connect the vent hose to the case. ☐ Task completed

19. Connect the speedometer cable to the transfer case. ☐ Task completed

20. Install and connect the rear driveshaft to the transfer case. ☐ Task completed

21. Install and connect the front driveshaft to the transfer case. ☐ Task completed

22. Connect all electrical terminals and wires according to the labels put on during removal of the transfer case. ☐ Task completed

23. Lower the vehicle. ☐ Task completed

24. Refill the transfer case with the correct fluid and amount. What type of fluid does this transfer case use?

____

____

Instructor's Comments

____

____

____

MANUAL TRANSMISSIONS JOB SHEET 36

Disassemble and Reassemble a Transfer Case

Name ______________________ Station ______________ Date ______________

NATEF Correlation

This Job Sheet addresses the following NATEF task:

E.4. Disassemble, service, and reassemble transfer case and components.

Objective

Upon completion of this job sheet, you will be able to disassemble, service, and reassemble a transfer case.

Tools and Materials

Pry bar

Torque wrench

Protective Clothing

Goggles or safety glasses with side shields

Describe the vehicle being worked on:

Year ________ Make ____________________ Model ____________________

VIN ____________________ Engine type and size ____________________

Transfer case type ____

PROCEDURE

1. If not previously drained, remove the drain plug and allow the oil to drain, then re-install the plug. ☐ Task completed
2. Loosen the flange nuts. ☐ Task completed
3. Remove the two output shaft yoke nuts, washers, rubber seals, and output yokes from the case. ☐ Task completed
4. Remove the four-wheel-drive indicator switch from the cover. ☐ Task completed
5. Remove the wires from the electronic shift harness connector. ☐ Task completed
6. Remove the speed sensor retaining bracket screw, bracket, and sensor. ☐ Task completed
7. Remove the bolts securing the electric shift motor and remove the motor. Note the location of the triangular shaft in the case and the triangular slot in the electric motor. ☐ Task completed
8. Loosen and remove the front case to rear case retaining bolts. ☐ Task completed

9. Separate the two halves by prying between the pry bosses on the case. ☐ Task completed
10. Remove the shift rail for the electric motor. ☐ Task completed
11. Pull the clutch coil off the main shaft. ☐ Task completed
12. Pull the 2WD/4WD shift fork and lock-up assembly off the main shaft. ☐ Task completed
13. Remove the chain, driven sprocket, and drive sprocket as a unit. ☐ Task completed
14. Remove the main shaft with the oil pump assembly. ☐ Task completed
15. Slip the high-low range shift fork out of the inside track of the shift cam. ☐ Task completed
16. Remove the high-low shift collar from the shift fork. ☐ Task completed
17. Unbolt and remove the planetary gear mounting plate from the case. ☐ Task completed
18. Pull the planetary gearset out of the mounting plate. ☐ Task completed
19. Install the input shaft and front output shaft bearings into the case. ☐ Task completed
20. Apply a thin bead of sealer around the ring gear housing. ☐ Task completed
21. Install the input shaft with planetary gear set and tighten the retaining bolts to specifications. What are the specifications?

    ____

    ____

22. Install the high-low shift collar into the shift fork. ☐ Task completed
23. Install the high-low shift assembly into the case. ☐ Task completed
24. Install the main shaft with oil pump assembly into the case. ☐ Task completed
25. Install the drive and driven sprockets and chain into position in the case. ☐ Task completed
26. Install the shift rails. ☐ Task completed
27. Install the 2WD/4WD shift fork and lock-up assembly onto the main shaft. ☐ Task completed
28. Install the clutch coil onto the main shaft. ☐ Task completed
29. Clean the mating surface of the case. ☐ Task completed
30. Position the shafts and tighten the case halves together. Tighten attaching bolts to specifications. What are the specifications?

    ____

    ____

31. Apply a thin bead of sealer to the mating surface of the electric shift motor. ☐ Task completed
32. Align the triangular shaft with the motor's triangular slot. ☐ Task completed
33. Install the motor over the shaft; wiggle the motor to ensure that it is fully seated on the shaft. ☐ Task completed

34. Tighten the motor's retaining bolts to specifications. What are the specifications?

____

____

35. Reinstall the wires into the connector and connect all electric sensors. ☐ Task completed

36. Install the companion flanges' seal, washer, and nut. Then tighten the nut to specifications. What are the specifications?

____

____

37. Summarize how you did.

____

____

____

____

Instructor's Comments

____

____

____

MANUAL TRANSMISSIONS JOB SHEET 37

Inspecting Front Wheel Bearings and Locking Hubs

Name ______________________ Station ______________ Date ________________

NATEF Correlation

This Job Sheet addresses the following NATEF task:

E.5. Inspect front wheel bearings and locking hubs; perform necessary action.

Objective

Upon completion of this job sheet, you will be able to inspect front wheel bearings and locking hubs on a 4WD vehicle.

Tools and Materials

Lift

Service manual

Protective Clothing

Goggles or safety glasses with side shields

Describe the vehicle being worked on:

Year _________ Make ____________________ Model ____________________

VIN ____________________ Engine type and size ________________________

PROCEDURE

1. Raise the vehicle to a height that allows you to work comfortably around the vehicle's wheels. ☐ Task completed
2. If the vehicle has locking hubs, rotate a front wheel slightly and move the hub selector to the "lock" position. Did the hub engaged with a click? ______________ ☐ Task completed
3. Does it feel like the axle is now rotating with the wheel? ☐ Task completed
4. Do the same thing to the wheel and hub on the other side of the vehicle. ☐ Task completed
5. Based on these checks, what are your conclusions about the condition of the hubs?

____

____

6. Grasp the front and rear of a front tire and push it in and out. Do you feel any endplay?

____

____

7. Do the same to other front tire. Did you feel any endplay?

____

____

8. Based on these two checks, do you think the wheel bearings need to be adjusted? Why or why not?

____

____

9. If adjustment is needed, remove the tire and wheel assembly. ☐ Task completed
10. Remove the snap ring at the end of the axle's spindle. ☐ Task completed
11. Remove the axle shaft spacers, thrust washers, and outer wheel bearing locknut. Keep the spacers and washers in the order they were so you can assemble the unit correctly. ☐ Task completed
12. Loosen the inner bearing locknut and then fully tighten it to seat the bearings. ☐ Task completed
13. Spin the brake rotor on its spindle. ☐ Task completed
14. Loosen the inner bearing locknut approximately one quarter-turn or the amount specified in the service manual. ☐ Task completed
15. Install the axle shaft spacers and washers, then install the outer bearing locknut. ☐ Task completed
16. Tighten the outer locknut to the specified torque. The torque spec is ____________.
17. Install a new snap ring into the end of the axle's spindle. ☐ Task completed
18. Repeat the wheel bearing adjustment procedure on the other front wheel. ☐ Task completed

Instructor's Comments

____

____

____

MANUAL TRANSMISSIONS JOB SHEET 38

Check Fluid in a Transfer Case

Name ______________________ Station ______________ Date ______________

NATEF Correlation

This Job Sheet addresses the following NATEF task:

E.6. Check drive assembly seals and vents; check lubricant level.

Objective

Upon completion of this job sheet, you will be able to inspect a transfer case for leaks and properly check its fluid level.

Tools and Materials

Hand tools

Service manual

Protective Clothing

Goggles or safety glasses with side shields

Describe the vehicle being worked on:

Year ________ Make ____________________ Model ____________________

VIN ____________________ Engine type and size ____________________

Describe the type of system and the model of the transfer case:

____

PROCEDURE

1. Raise and support the vehicle. ☐ Task completed
2. Carefully inspect the area around the transfer case for signs of oil leakage. Summarize your findings. ☐ Task completed

   ____

   ____

3. Check the connections for the driveshafts to the transfer case. The presence of oil may indicate bad seals. Summarize what you found.

____

____

4. Check the mounting point for the transfer case at the transmission, looking for signs of fluid that may indicate a gasket problem at this connection. Summarize what you found.

____

____

5. Locate the fill plug on the transfer case. (Refer to the service manual for the location of the plug.) ☐ Task completed

6. Remove the filler plug. ☐ Task completed

7. Using your little finger, feel in the hole to determine if you can touch the fluid. ☐ Task completed

8. If you cannot touch the fluid, refer to the service manual for fluid type. Fill the transfer case. The recommended fluid is ____________________. ☐ Task completed

9. If the transfer case is low on fluid, visually inspect it to locate the leaks. ☐ Task completed

10. If you can reach the fluid with your finger, check the smell, color, and texture of the fluid. ☐ Task completed

11. If the fluid is contaminated, determine what is contaminating it. Then correct that problem and drain the fluid from the transfer case and refill it with clean fluid. ☐ Task completed

Problems Encountered

____

____

____

Instructor's Comments

____

____

____

MANUAL TRANSMISSIONS JOB SHEET 39

Servicing 4WD Electrical Systems

Name ______________________ Station ______________ Date ________________

NATEF Correlation

This Job Sheet addresses the following NATEF task:

E.7. Diagnose test, adjust, and replace electrical and electronic components of four-wheel-drive systems.

Objective

Upon completion of this job sheet, you will be able to diagnose test, adjust, and replace the electrical and electronic components of four-wheel-drive systems.

Tools and Materials

Component locator
Small mirror
Flashlight
Scan tool
DMM
Wiring diagram
Air nozzle
Service manual

Protective Clothing

Goggles or safety glasses with side shields

Describe the vehicle being worked on:

Year ________ Make ____________________ Model ___________________

VIN ____________________ Engine type and size _______________________

PROCEDURE

1. Check all electrical connections to the transfer case. Make sure they are tight and not damaged. Record your findings.

____

____

2. Release the locking tabs of the connectors and disconnect them, one at a time, from the transfer case. Carefully examine them for signs of corrosion, distortion, moisture, and transmission fluid. A connector or wiring harness may deteriorate if automatic transmission fluid reaches it. Using a small mirror and flashlight may help you get a good look at the inside of the connectors. Record your findings.

____

____

3. Inspect the entire wiring harness for tears and other damage. Record your findings.

____

____

4. Because the operation of the engine, transmission, and transfer case are integrated through the control computer, a faulty engine sensor or connector may affect the operation of all of these. With a scan tool, retrieve any diagnostic trouble codes (DTCs) saved in the computer's memory. Were there any present? If so, what is indicated by them?

____

____

5. The engine control sensors that are the most likely to cause shifting problems are the throttle-position sensor, manifold absolute position (MAP) sensor, and vehicle speed sensor. Locate these sensors and describe their location.

____

____

____

6. Remove the electrical connector from the throttle-position (TP) sensor and inspect both ends for signs of corrosion and damage. Record your findings.

____

____

7. Inspect the wiring harness to the TP sensor for evidence of damage. Record your findings.

____

____

8. Check both ends of the three-pronged connector and wiring at the MAP sensor for corrosion and damage. Record your findings.

____

____

9. Check the condition of the vacuum hose. Record your findings.

____

____

10. Check the speed sensor's connections and wiring for signs of damage and corrosion. Record your findings.

____

____

11. Record your conclusions from the visual inspection.

____

____

____

____

12. Remove the selector switch and check its action with an ohmmeter. Move the switch to all possible positions and observe the ohmmeter. Compare the action of the switch to the circuit shown in the wiring diagram. Summarize the condition of the switch.

____

____

____

13. Locate the transfer case shift motor or solenoid. ☐ Task completed

14. Carefully inspect this for damage and poor mounting. Record your findings here.

____

____

15. Check the solenoid or motor following the procedures given by the manufacturer. Summarize this check and your results.

____

____

16. If the solenoid or motor must be removed for testing, make sure it is mounted correctly and the mounting bolts tighten to specs when you reinstall it. ☐ Task completed

17. Check the service manual and identify all switches located on the transfer case. List these here.

____

____

____

18. Determine the function of each switch. Is it a grounding switch or does it complete or open a power circuit?

____

____

____

19. An ohmmeter can be used to identify the type of switch being used and to test the operation of the switch. There should be continuity when the switch is closed and no continuity when the switch is open. Record your findings.

____

____

20. Apply air pressure to the part of the switch that would normally be exposed to oil pressure and check for leaks. Record your findings.

____

____

21. The transfer case may be equipped with a speed sensor. Typically the speed sensor is a permanent magnetic (PM) generator. Locate the speed sensor and describe its location.

____

____

22. With the vehicle raised on a lift, allow the wheels to be suspended and free to rotate. ☐ Task completed

23. Set your DMM to measure AC voltage. ☐ Task completed

24. Connect the meter to the speed sensor. ☐ Task completed

25. Start the engine and put the transmission in gear. Slowly increase the engine's speed until the vehicle is at approximately 20 mph, and then measure the voltage at the speed sensor. Record your findings.

____

____

26. Slowly increase the engine's speed and observe the voltmeter. The voltage should increase smoothly and precisely with an increase in speed. Record your findings.

____

____

27. A speed sensor can also be tested with it out of the vehicle. Connect an ohmmeter across the sensor's terminals. What is the measured resistance?

____

____

28. Locate the specifications for the sensor and compare your readings with specifications.

____

____

Instructor's Comments

____

____

____